Impact of Technological Innovation on the Poor

DEVELOPMENT ECONOMICS AND POLICY

Series edited by Franz Heidhues †, Joachim von Braun,
Ulrike Grote and Manfred Zeller

Vol. 76

Impact of Technological Innovation on the Poor

Integrated Aquaculture-Agriculture in Bangladesh

Abu Hayat Md. Saiful Islam

Bibliographic Information published by the Deutsche Nationalbibliothek
The Deutsche Nationalbibliothek lists this publication in the Deutsche
Nationalbibliografie; detailed bibliographic data is available in the internet at
http://dnb.d-nb.de.

Zugl.: Bonn, Univ., Diss., 2015

Library of Congress Cataloging-in-Publication Data
Names: Islam, Abu Hayat Md. Saiful, 1982- author.
Title: Impact of technological innovation on the poor : integrated aquaculture-
agriculture in Bangladesh / Abu Hayat Md. Saiful Islam.
Other titles: Integrated aquaculture-agriculture in Bangladesh | Development
economics and policy ; Bd. 76.
Description: First edition. | Frankfurt am Main ; New York : Peter Lang, 2016. |
Series: Development economics and policy, ISSN 0948-1338 ; vol. 76
| Includes bibliographical references and index.
Identifiers: LCCN 2016001543 | ISBN 9783631671412
Subjects: LCSH: Integrated agricultural systems—Economic aspects—
Bangladesh. | Integrated aquaculture—Economic aspects—Bangladesh.
Classification: LCC S471.B3 I85 2016 | DDC 338.1095492—dc23 LC record
available at http://lccn.loc.gov/2016001543

D 5
ISSN 0948-1338
ISBN 978-3-631-67141-2 (Print)
E-ISBN 978-3-653-06493-3 (E-Book)
DOI 10.3726/978-3-653-06493-3

© Peter Lang GmbH
Internationaler Verlag der Wissenschaften
Frankfurt am Main 2016
All rights reserved.
PL Academic Research is an Imprint of Peter Lang GmbH.

Peter Lang – Frankfurt am Main · Bern · Bruxelles · New York ·
Oxford · Warszawa · Wien

All parts of this publication are protected by copyright. Any
utilisation outside the strict limits of the copyright law, without
the permission of the publisher, is forbidden and liable to
prosecution. This applies in particular to reproductions,
translations, microfilming, and storage and processing in
electronic retrieval systems.

This publication has been peer reviewed.

www.peterlang.com

Table of Contents

Table of Contents V

List of Tables IX

List of Figures XI

List of Abbreviations XIII

Acknowledgements XV

Chapter 1: Introduction 1
1.1 Background 1
1.2 Problem Statement 2
1.3 Research Objectives and Questions 5
1.4 Conceptual Framework 6
1.5 Research Methods 8
    1.5.1 Study Area 8
    1.5.2 Data: Sampling Technique, Sample Size, and Survey Effort 10
    1.5.3 Indigenous People of Bangladesh 12
1.6 Outline of the Dissertation 13

Chapter 2: Performance of Integrated Aquaculture-Agriculture Value Chain Development in Bangladesh 15
2.1 Introduction 15
2.2 Methods 17
    2.2.1 Data 17
    2.2.2 Analytical Methods 18
        2.2.2.1 Value-chain analysis 18
        2.2.2.2 Gross margin analysis 20
        2.2.2.3 Partial budgeting analysis 21
        2.2.2.4 SWOT analysis 21

2.3 Results and Discussion 22
    2.3.1 Value Chain Mapping 22
    2.3.2 Actors, Value Addition, Governance, Institutional Framework and Employment in Rice–fish based Integrated Aquaculture-agriculture Value Chains 24
    2.3.3 Gross Margin Analysis of Value Chain Actors 28
    2.3.4 Partial Budgeting 32
    2.3.5 SWOT Analysis of Integrated Rice–fish Value Chains 33
        2.3.5.1 Strengths 36
        2.3.5.2 Weaknesses 37
        2.3.5.3 Opportunities 38
        2.3.5.4 Threats 40
2.4 Conclusions and Policy Implications 42

Chapter 3: Integrated Aquaculture-Agriculture Value Chain Participation Dynamics in Bangladesh 45

3.1 Introduction 45
3.2 Integrated Aquaculture-Agriculture Value Chain Participation Dynamics: Issues and Approaches 48
    3.2.1 Integrated Aquaculture-Agriculture in Bangladesh 48
    3.2.2 Integrated Aquaculture-Agriculture Adoption Research Issues 49
3.3 Empirical Econometric Estimation Framework 50
    3.3.1 Conceptual Framework for Integrated Aquaculture-Agriculture Value Chain Participation 50
        3.3.1.1 Multinomial logit model 52
        3.3.1.2 Random Effects logit model 53
        3.3.1.3 Re-specified Random Effects logit model for addressing endogeneity 53
        3.3.1.4 Correlated Random Effects model 54
3.4 Data and Descriptive Statistics 54
3.5 Econometric Results 58
    3.5.1 Multinomial Logit Model Results 58
    3.5.2 Random Effects Logit Model Results 60

3.6 Conclusions and Policy Implications 63

Chapter 4: Welfare Impacts of Integrated Aquaculture-Agriculture Value Chain Participation Dynamics in Bangladesh 67

4.1 Introduction..67
4.2 Framework of the Study..70
4.3 Literature Review 71
4.4 Data and Descriptive Statistics..73
 4.4.1 Who Participated in Integrated Aquaculture-Agriculture Value Chains in Bangladesh?...73
 4.4.2 Relationships between Integrated Aquaculture-Agriculture Value Chain Participation Dynamics and Household Welfare..........75
 4.4.3 Distributional Impacts of Integrated Aquaculture-Agriculture Value Chain Participation..76
4.5 Estimation Issues and Strategy..77
4.6 Results..81
 4.6.1 Integrated Aquaculture-Agriculture Value Chain Participation Dynamics and Household Welfare 81
  4.6.1.1 Naive Pooled Ordinary Least Squares and Random Effects model results 81
  4.6.1.2 Standard Fixed Effects model results............................ 83
  4.6.1.3 Heckit panel and control function approach results 85
 4.6.2 Who Benefits More From Integrated Aquaculture-Agriculture Value Chain Participation? 86
4.7 Conclusions 87

Chapter 5: Comparative Socio-environmental Impacts of Rice–Fish Based Integrated Aquaculture-Agriculture and Rice Monoculture in Bangladesh 91

5.1 Introduction..91
5.2 Methods..94
 5.2.1 Data...94

VII

     5.2.2 Econometric framework 95
     5.2.3 Dependent Variables: Farmer Socio-Environmental
           Awareness Index 98
     5.2.4 Independent Variables 98
5.3 Results and Discussion 100
     5.3.1 Plot Level Comparison of Inputs Used for Rice–fish based
           Integrated Aquaculture-Agriculture and Rice Monoculture
           Farming Systems 100
     5.3.2 Comparison of Farmer Perceptions 105
           5.3.2.1 Rice–fish based IAA farmer perceptions of the socio-
                   environmental impacts of rice monoculture and
                   integrated rice–fish systems 105
           5.3.2.2 Comparison of perceptions of the socio-environmental
                   impacts of rice monoculture among integrated rice–
                   fish farmers and rice monoculture farmers 108
     5.3.3 Determinants of Farmer Awareness of the Socio-
           environmental Impacts of Rice Monoculture: Tobit and PSM
           Analyses 110
5.4 Conclusions and Policy Implications 112

Chapter 6: Summary, Conclusion—Policy Implications and Further Research Needs 117

6.1 Research Summary 117
6.2 Conclusions and Policy Implications 123
6.3 Further Research Needs 125

References 127

Appendix 161

Abstract 195

Zusammenfassung 197

List of Tables

Table 1.1:	Sample size of the panel survey of integrated aquaculture-agriculture participation among study area households in Bangladesh	11
Table 2.1:	Sample sizes by rice–fish based integrated aquaculture-agriculture value chain participation category in Bangladesh	18
Table 2.2:	Labour allocation patterns (person days/hectare/year) for rice monoculture and integrated rice–fish production systems in Bangladesh	27
Table 2.3:	Mean quantity and cost/return values of inputs and outputs of different farming systems in Bangladesh	29
Table 2.4:	Mean costs/returns of inputs and outputs of actors along integrated rice–fish farming system fish value chains in Bangladesh	31
Table 2.5:	Partial budgeting analysis results: net changes in gross margins are due to the replacement of rice monoculture with integrated rice–fish farming systems in Bangladesh	33
Table 2.6:	SWOT framework results for integrated rice–fish farming system value chain development in Bangladesh.	33
Table 3.1:	Mean and standard deviations of independent variables by integrated aquaculture-agriculture value chain participation category in Bangladesh	56
Table 3.2:	Reported reasons for dis-participation from integrated aquaculture-agriculture value chains in Bangladesh	58
Table 3.3:	Multinomial logit analysis results of integrated aquaculture-agriculture value chain participation in Bangladesh	59
Table 3.4:	Random Effects model results of IAA value chain participation	62
Table 4.1:	Descriptive statistics (mean and standard deviation) of Integrated Aquaculture-Agriculture participation explanatory variables in Bangladesh	74
Table 4.2:	Pooled Ordinary Least Squares and Random Effects model results	82

Table 4.3:	Fixed Effects model results of the relationship between IAA value chain participation and household income and consumption frequency of selected foods in Bangladesh	84
Table 4.4:	Heckit panel and control function model coefficients of the relationship between IAA participation and household income in Bangladesh	85
Table 4.5:	Comparison of the Fixed Effects model results of annual household income* for the IAA value chain actors by relative wealth in Bangladesh	86
Table 5.1:	Summary statistics of independent variables used in the regression	99
Table 5.2:	Comparative farm level socio-environmental benefits and costs of rice–fish based integrated aquaculture-agriculture and rice monoculture production	102
Table 5.3:	Rice–fish based IAA farmer perceptions of the impacts of rice monoculture and integrated rice–fish systems in Bangladesh	107
Table 5.4:	Rice–fish based IAA system and rice monoculture farmer perceptions on the socio-environmental impacts of rice monoculture systems in Bangladesh	109
Table 5.5:	Tobit model results for determinants of farmer perceptions of the socio-environmental impacts of rice monoculture in Bangladesh	111
Table 5.6:	ATT results from the alternative matching algorithm	112
Table A.1.1:	Attrition bias test results	161
Table A.3.1:	Mean and standard deviations (in parentheses) of independent variables by IAA value chain participation category	162
Table A.3.2:	Multinomial logit analysis of IAA value chain participation with aggregated sample	164
Table A.5.1:	Construction procedure of the socio-environmental awareness index	165
Table A.4.1:	Household well-being measures for IAA value chain participators, non-participators and dis-participators in Bangladesh	166
Table A.4.2:	Welfare distribution among the IAA value chain participation categories	167
Table A.1.2:	Interview schedule for IAA value chain actors in Bangladesh	168

List of Figures

Figure 1.1: Conceptual framework of the study 7
Figure 1.2: Map of the study area indicating the geopolitical districts (purple) and sub-districts (green) in Bangladesh 9
Figure 2.1: Schematic representation of rice–fish value chains in Bangladesh..23
Figure 4.1: Conceptual framework showing the welfare effect pathways of IAA value chain participation [Adapted after modification from von Braun (1988) and Dey et al. (2010)] 71
Figure 4.2: Household income by integrated aquaculture-agriculture value chain participation categories in Bangladesh..........................75
Figure 4.3: Distribution of income effects among integrated aquaculture-agriculture value chain participation categories in Bangladesh 76

List of Abbreviations

ADB	Asian Development Bank
AFP	*Adivasi* Fisheries Project
ATT	Average treatment effects on treated
BARC	Bangladesh Agricultural Research Council
BAU	Bangladesh Agricultural University
BBS	Bangladesh Bureau of Statistics
BCR	Benefit cost-ratio
BDT	Bangladeshi Taka
BER	Bangladesh Economic Review
BFRI	Bangladesh Fisheries Research Institute
BRKB	Bangladesh Rice Knowledge Bank
BRRI	Bangladesh Rice Research Institute
BW	Band width
CARE	Cooperative for Assistance and Relief Everywhere
CBO	Community-based organisation
CHT	Chittagong Hill Tracts
CRE	Correlated Random Effect
CSA	Climate smart agriculture
DEA	Department of Agricultural Extension
DOF	Department of Fisheries
DSAP	Development of Sustainable Aquaculture Project
EU	European Union
FAO	Food and Agricultural Organisation of the United Nations
FE	Fixed Effects
FFS	Farmer field school
FRSS	Fisheries Resources Survey System
GDP	Gross domestic product
GR	Green revolution
GHG	Greenhouse gas
GM	Gross margin
GOLDA	Greater options for local development through aquaculture
GR	Green Revolution
GWP	Global warming potential
HYV	High yield varieties
IAA	Integrated aquaculture-agriculture

IFAD	International Fund for Agricultural Development
IFPRI	International Food Policy Research Institute
IPM	Integrated pest management
IRFFS	Integrated rice–fish farming system
LDCs	Least developed countries
MOA	Ministry of Agriculture
MV	Modern varieties
NARS	National Agricultural Research System
NGOs	Non-governmental organisations
NN	Nearest neighbour
NOPM	New options for pest management
POLS	Pooled Ordinary Least Square
PRSP	Poverty Reduction Strategy Paper
PSM	Propensity score matching
RE	Random Effects
RF	Rice–fish culture
RF-IAA	Rice–fish based integrated aquaculture-agriculture
RM	Rice monoculture
SWOT	Strength, weakness, opportunities and threats
TC	Total cost
TOA-MD	Trade-off Analysis for multi-dimensional impact assessment
TVC	Total variable cost
VIF	Variance inflation factor
WDI	World development indicator
WF	WorldFish
WFP	World Food Program

Acknowledgements

This dissertation research has its origins in my life long captivation with rice and fish that began when I was a young child catching fish in the rice fields with my younger brother. Later in life this fascination was rekindled during my study of agricultural economics at Bangladesh Agricultural University (BAU), where I was drawn to the patterns of rice and fish sector development and transformation that I had observed unfolding in Asia, particularly in Bangladesh.

Many people and organisations have helped during the three-and-a-half year journey of my PhD research. Therefore, I would like express my deepest gratitude to all of those who have contributed to this dissertation, both in the academic sense and for heartening the fabric of my life abroad during the time spent on this process.

I very gratefully acknowledge the DAAD programme of the German Federal Ministry of Education and Research (BMBF) for funding my PhD research and the Dr Hermann Eiselen Doctoral Programme of the *Fiat Panis* foundation and the TIGA project for financing the field research and participation in several conferences. I am extremely grateful to my respected supervisor Prof. Dr Joachim von Braun for his invaluable advice and ever-incisive comments, unfailing support, untiring assistance, and his wisdom in every aspect of my work, from the very beginning to the completion of this research effort, and for pushing me to look further. I would like to pay special tribute to his affection, co-operation and coordination during the different stages and activities during the course of this dissertation work. I could not have wished for a better supervisor. I am also highly grateful to Prof. Dr Michael Frei for assuming the role of second supervisor and providing very useful comments on my preliminary dissertation draft. My appreciation also goes to Dr Franz W. Gatzweiler for his excellent tutoring on my work throughout the PhD process. I would like to thank Dr Kerstin K. Zander (ZEF alumni, now senior research fellow at the Northern Institute, Charles Darwin University, Australia) for her review and suggestions on Chapter 5, as well as other related documents. I also thank Dr Julia Anna Matz (ZEF senior researcher) for her comments on the econometric analyses.

My gratitude extends to WorldFish, especially the WorldFish Bangladesh office for providing the first two-waves of panel data (2007 and 2009), thereby enabling me to construct the three- wave of the panel dataset, and particularly to Dr Benoy K. Barman and Dr Khondker Murshed-e-Jahan from the WorldFish Bangladesh office, and to Dr Charles Crissman from the WorldFish main office in Malaysia for their generous cooperation and guidance, which was essential for my ability to negotiate my way through the field research. I would also like to express my sincere thanks to the many farmers and other value chain participants who were kind enough take the time to answer my long lists of numerous questions in Bangladesh, and for their openness and hospitality that I often received during the interviews. I am also very much grateful to my respected teachers and colleagues at the department of Agricultural Economics at the Bangladesh Agricultural University (BAU) and to the BAU authority for granting me a study leave to pursue my higher studies at the University of Bonn.

My sincere appreciation goes to Dr Gunther Manske, Dr Guido Luechters, Mrs Maike Retat-Amin, Mrs Rosemarie Zabel, Dr Samantha Antonini, and all of my other friends and colleagues at ZEF, especially my cohorts and officemates; Justice, Lukas and Elias, and the other Bangladeshi students and community living in Bonn for helping me with my research and making my PhD life more exciting and enjoyable.

My sincere appreciation and indebtedness also go to my beloved wife, Fahmida Bhuyan, and to my parents, whose affection, inspiration, sacrifice, encouragement and continuous blessing were a great moral support and inspiration for my work.

Above all, I do not have the words to express my gratefulness to Allah Almighty, who gave me enough strength, courage and wisdom, and whose ultimate design and mercy made my dreams a reality.

Chapter 1: Introduction

1.1 Background

Rice and fish are integral to society in many Asian countries where they provide the basis of food security and well-being. Integrated rice–fish farming systems (IRFSS) in this region are quite varied. A broad spectrum of integrated aquaculture-agriculture (IAA) systems have been practiced for centuries in Asia, particularly in Bangladesh, China, India, Indonesia, Malaysia, Thailand, and Vietnam (Prein, 2002; Dey et al., 2013). Many of these IAA systems, especially rice–fish based IAA systems, have been transformed during the course of the green revolution (GR) in Asia due to the unsustainable intensification of rice production through intensive use of fertilisers, pesticides, and irrigation. With respect to renewed interest in sustainable intensification under a paradigm of a 'doubly green revolution,' IAA systems can be a potential intensification strategy. Through participatory research and extension systems, different national and international governmental and non-governmental organisations (NGOs) together with innovative farmers are making an effort to address the unintended consequences of GR intensification efforts. Many actors have introduced improved IAA systems in recent decades that are suited to the geographically specific farming environments and resource endowments of local farmers. One recent estimate indicates that about 0.18 million hectares of land are currently under rice–fish based IAA systems in Bangladesh, which is much lower than the potentially suitable area of 2 to 3 million hectares (ADB, 2005; Dey et al., 2013). This raises the question of whether or not the adoption and impacts of IAA systems have been adequately examined. A recent meta-review of rice–fish based IAA systems in Bangladesh also indicates that relevant socio-economic research is relatively scarce considering the potential of these systems for improving agricultural production and rural livelihoods in Bangladesh (Dey et al., 2013). Troell et al. (2014; 13257) stated that "interconnections between the aquaculture, crop, livestock, and fisheries sectors can act as an opportunity for, enhanced resilience in the global food system given the increased resource scarcity and climate change and if government policies provide adequate incentives for resource efficiency, equity, and environmental protection."

Given this backdrop we used the value chain conceptual framework as the basis to investigate the financial performance of IAA value chain actors and the dynamics and determinants of IAA value chain participation, and the welfare and environmental impacts among extremely poor and marginalized indigenous[1] populations in Bangladesh. The intent of the research presented here was to better understand how the current nature of IAA system development acts as either an impediment to, or an opportunity for, the enhancement of smallholder welfare given their typically limited resource endowments and numerous constraints. To accomplish this we considered all of the IAA value chain actors and system level research methods in an integrated manner rather than using commodity specific (e.g. only rice or only fish) or technology specific approaches (e.g. improved seed, fertiliser, irrigation). In doing so the research effort is expected to contribute to the on-going debate on sustainable intensification policy and practices within the agricultural sector, and to be relevant to researchers on these systems and related systemic challenges in other sectors and regions.

1.2 Problem Statement

Despite immense progress in poverty reduction in the developing world there will continue to be around one billion people living below the international poverty level of US$1.25 per person per day in 2015 and 162 million people who live in 'ultra-poverty' (less than US$0.50 per person per day). Many people living under the US$1.25 poverty line are vulnerable to poverty and food insecurity (Ahmed et al., 2007; Chen and Ravallion, 2012). Similarly, more than one billion people in the world are chronically undernourished and most of these live in Asia and the Pacific (FAO, 2010a). The common characteristics of the world's poorest and most hunger prone people are that they typically reside in rural areas that are remote with respect to access to roads, markets, schools, and health services, and they are less likely to be educated and more likely to belong to socio-ethnic minorities and other marginalised groups (Ahmed et al., 2007). Food security and poverty reduction continue to be daunting challenges for most developing countries, however, the pressing question is how can both food insecurity and poverty be reduced?

Many of the poor and vulnerable rely heavily on the agricultural sector. Although the GDP share of agriculture is declining in most countries, it continues to be the backbone of the economies of most of the least developed countries (LDCs). It is the largest and most significant source of livelihoods in terms of providing food, income, and employment in LDCs. It produces a multiplier effect through strong forward and backward linkages with other sectors and an added stimulus for growth and income generation (Mellor, 1998). Agriculture is also one of the

---

1 The terms 'indigenous,' '*adivasi*,' 'aboriginal,' 'ethnic minority,' and 'tribal' are used synonymously throughout the dissertation.

major sources of economic development and recovery. Agricultural growth has powerful impacts on poverty via its broad effects on poor people, which may not be the case for growth in the manufacturing and service sectors. Thus, sustaining the productivity and efficiency of the agricultural sector is the central emphasis for 'pro-poor' growth with respect to economic planning (Thirtle et al., 2003; Koroma, 2007).

Agriculture sector is typically dominated by crop production, especially of rice in Asia. Although rice production has increased substantially since the onset of the GR, due to rising food demand it is estimated that production needs to increase by more than 50% over the next few decades (Spielman and Pandya-Lorch, 2009; Mishra and Salokhe, 2010). Moreover, current concerns about the environment and food security, including food safety, are gaining momentum, which feed the debate about the sustainability of GR approaches in developing countries (Redclift, 1989; Alauddin and Tisdell, 1991; Shiva, 1991; Singh, 2000; Kunio, 2002). Simultaneously, increases in global commodity prices due to new drivers like increases in the demand for feed, food, and biofuels are putting significant pressure on agricultural systems (von Braun, 2007). An important concern with respect to feeding growing populations is whether or not existing rice research systems will be able to sustain the growth of rice production or if new solutions are required to sustainably meet the demand for rice by a growing world population (Surridge, 2004).

Horizontal expansion of arable land is not possible in many areas and in some cases it is declining, so the only possible way to increase the productivity of land, labour and water resources is vertical intensification through the integration of different agricultural enterprises or by changing management practices and efficiency through sustainable intensification and resource reallocation. Such integration efforts could reduce poverty and malnutrition. Accordingly, 'doubly green revolution'[2] perspectives call for innovative strategies that are both 'pro-poor' and technically feasible, addressing livelihoods in an economically, socially, and environmentally acceptable way, which has gained much attention in recent literature

---

2   'Doubly green revolution' not only signifies productive, but also stable, resilient, and equitable means of providing benefits to everyone. This signifies that it should be equitable, sustainable, and environmentally benign (Conway, 2011). Conway argues for a 'doubly green revolution' that is characterised by sustainable productivity and conservation. He proposes to emphasise the design or development (or rediscovery) of improved crop and livestock varieties, alternatives to inorganic fertilisers and pesticides, improvements to soil and water management practices, and enhancement of earning opportunities for the poor, especially women, through interaction between researchers and farmers (Conway, 1999).

in the context of poverty reduction and sustainable agricultural development in developing countries (Conway, 1999; Noltze et al., 2011). IAA farming systems could provide such a tool for increasing carbohydrate and protein production more sustainably by using scarce land and water resources in an intensive and complementary way (Meaden and Kapetsky, 1991). IAA offers the prospect of higher rice yields with more efficient agricultural land use with limited negative impacts on natural resources at affordable costs for poor smallholder farmers (Khoo and Tan, 1980; Ruddle, 1982; Lightfoot et al., 1992; Frei and Becker, 2005a). Edwards (2000) mentioned that there is significant potential for increased involvement of poor farming households in rice–fish production with respect to both rain fed and irrigated systems, and mentioned that there are many successful examples in Bangladesh, Madagascar, and Thailand. Several studies in different countries have identified the advantages of IAA in terms of more efficient use of land and water resources, increased food production, greater food and nutritional security, improved farmer income (Mukherjee, 1995; Gupta et al., 1996, 1998; Purba, 1998; Horstkotte-Wesseler, 1999; Berg, 2002; Ahmed et al., 2008; Nahar, 2010; Ahmed and Garnett, 2011; Rahman et al., 2011), and for the control of rice weeds, pests, and mosquitoes (Neng et al., 1995; Rothuis et al., 1998a; Vromant et al., 1998; Berg, 2001; Ichinose et al., 2002; Frei and Becker, 2005b). In spite of these immense benefits and the research and promotional efforts of many international and national organisations, IAA farming has not been widely adopted in Asia (Rothuis, 1998a; Ahmed et al., 2008). This issue elicits the question of whether or not the adoption of IAA and its impacts are adequately understood.

Most of the published and unpublished research on IAA farming is incomplete. There is a general lack of high quality detailed field research using scientific and proper econometric methods. To date there is only limited published research on the economics of rice–fish–prawn culture (Ali and Mateo, 2007). There is a lack of applied research using proper econometric methods and theory on adoption patterns and intensity, and of the impacts of IAA on poverty, food security, gender, and land, labour, credit markets, the environment, income diversification, and equity, especially in terms of extremely poor and marginalised populations. Research on rice–fish based IAA has focused primarily on biological and technical issues that are location and season specific rather than at the system level or across entire annual agricultural cycles. Socio-economic, policy, and institutional dimensions of rice–fish based IAA system research is generally lacking (Dey et al., 2013).

So there is a need to systematically examine the above mentioned issues before widespread diffusion of IAA farming in potential implementation areas of the world, including Bangladesh. For more widespread diffusion and poverty reducing policy intervention it is necessary to have a better understanding of adoption patterns, as well as of the impacts of the adoption of new technologies in terms of household welfare (Becerril and Abdulai, 2010; Noltze et al., 2012). Similarly, Feder and Umali (1993) mentioned that for the determination of cost-effective policy options and their optimal intensity and duration it is necessary to know whether or not technology is adopted in packages, individually, in combination with other factors, or following a sequence. Feder et al. (1985) (in Doss, 2006; 208) mentioned that future technology adoption research is primarily needed in the following five areas: (i) the intensity of adoption (not just dichotomous choices); (ii) the simultaneity of adoption of different components of a technology package; (iii) the impacts of incomplete markets and policies on adoption decisions; (iv) the contextualization of adoption decisions within social, cultural and institutional environments; and (v) the dynamic patterns of land tenure and wealth accumulation among early and late stage IAA adopters. Doss (2006; 208) mentioned that, 20 years after Feder et al. technology adoption studies have made substantial progress, especially in the first two areas and that "[t]he issues of how institutional and policy environments affect the adoption of new technologies and how the dynamic patterns of adoption affect the distribution of wealth and income remain unanswered." Given this backdrop, this study attempted to fill research gaps mentioned above by using more sophisticated analytical approaches based on an integrated framework under a broader range of geographical and institutional conditions in marginalised, extreme poverty settings in Bangladesh. Thus it is expected that the study will facilitate the ability of policy-makers and international development organisations to make more nuanced decisions about the optimal entry point for addressing rural poverty and assessing approaches to rural development and their effectiveness in reducing rural poverty in developing countries.

1.3 Research Objectives and Questions

The study examined causality among factors that affect IAA value chain participation and its impacts in terms of welfare and the environment. Externalities of system changes were explicitly modelled (i.e. welfare benefits, environmental benefits or costs). Specifically, the study sought to:

1. Evaluate the overall performance of IAA value chain development in Bangladesh,
2. Determine the factors that influence the dynamics of participation in IAA value chains,

3. Assess the welfare impacts of IAA value chain participation dynamics,
4. Analyse the competitiveness and environmental impacts of rice–fish based IAA relative to rice monoculture farming systems.

Based on the stated objectives and research problems, the study sought to answer the following research questions:

1. Is participation in IAA value chains a profitable option and how can overall performance be improved?
2. What are the factors affecting the decision of whether or not to participate in IAA value chains and how do these factors differ among the participator categories?
3. How do IAA value chain participation dynamics affect the welfare of the marginalized rural poor?
4. How do farmers perceive the environmental effects of rice–fish based IAA diffusion relative to rice monoculture and the factors determining such perceptions?

1.4 Conceptual Framework

The conceptual framework briefly summarizes the key assumptions, hypotheses and research questions of the study on three levels of analysis (Figure 1.1). The first level is the individual value chain analysis, which indicates the key actors in a chain, estimates the value added at each stage, establishes the chain dynamics and governance, as well as financial performance of each actor. The second level is the value chain participation dynamics. This level focuses mainly on socio-economic, institutional, and physical/natural environmental contexts. The aim at this level is to observe how these arrangements facilitate or hinder participation and performance of the different value chain actors. The third level concerns the impacts of value chain participation dynamics through production intensification, livelihood strategy, and change over time. Here the emphasis was to reveal two levels of impact; the intended impact (i.e. socio-economic welfare impacts) and unintended (externality) impacts (i.e. environmental impacts). In addition, these impacts, along with physical, socio-economic and institutional factors, may work as a feedback mechanism that helps to explain participation dynamics. The impacts of this system can act equally as an incentive or a disincentive to drive or divert further livelihood strategies through IAA value chain participation. These three levels of analysis will provide greater insight, not only to what is happening with respect to IAA in Bangladesh, but why it is happening and what needs to be done to improve the situation. The conceptual framework is reflected by the study objective to take into account the socio-economic and environmental impacts (including positive and negative externalities) associated with system change (i.e. from rice/fish monoculture to IAA system). To capture all the positive and negative externalities, a comprehensive social and economic assessment is required, but we capture some part of it and the rest needs further research (see further research need details in Chapter 6, section 6.3).

Figure 1.1: Conceptual framework of the study

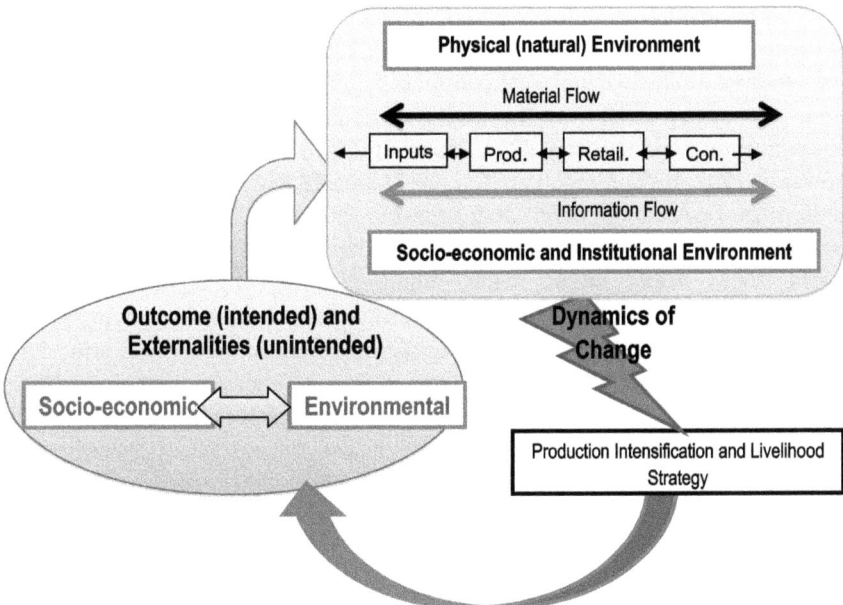

The conceptual framework highlights IAA as the key element of this study. At this stage it is necessary to conceptualise IAA, as it will be featured throughout the dissertation. Asia is commonly considered to be the origin of IAA systems, which were initially developed in China as a means of increasing food production on small-scale subsistence farms with limited resources (Edwards, 2003; Little and Muir, 2003). IAA is based on the concept of integrated resource management where "an output from one sub-system in an integrated farming system which may otherwise may have been wasted becomes an input to another sub-system resulting in a greater efficiency of output of desired products from the land/water area under the farmer's control" (Edwards et al., 1988; 5). Similarly Prein (2002; 128) defined this system as "concurrent or sequential linkages between two or more human activity systems, one or more of which is aquaculture, directly on-site, or indirectly through off-site needs and opportunities, or both." More broadly it is the linkages between two or more farming activities, of which at least one is aquaculture (Edwards, 1987). Generally there are two forms of IAA, one of which is pond-based IAA, in which a pond enterprise is added to a farm system (i.e. appropriate fish species stocked in a pond and available input materials from the farm such as crop and livestock residues are used as fish feed) (Brummett and Noble, 1995; Prein, 2002). The other form is where an aquaculture system is physically integrated into another system through redesign and re-operationalization of the latter, (e.g. rice–fish based IAA) (Prein, 2002). Both types of IAA systems and other value chain actors were considered in this study.

In IAA synergies between farming systems increase productivity, efficiency, diversification,[3] and sustainability (Talpaz and Tsur, 1982; Edwards et al., 1988; Edwards, 1989, 1998; Alsagoff et al., 1990; Dalsgaard and Oficial, 1997; Gomiero et al., 1999; DSAP, 2005; Berg, 2002; Jamu and Piedrahita, 2002; Frei and Becker, 2005a; Pant et al., 2005; Nhan et al., 2007; Tipraqsa et al., 2007; Jahan et al., 2008; Blythe, 2013), reduce environmental pollution by recycling aquaculture wastes (solids, organics and nutrients) and farm nutrient loss as well as by utilizing wastes produced from agriculture as feeds or fertilisers for aquaculture (Little and Muir, 1987; Edwards, 1998, 1993; Costa-Pierce, 2002; Devendra and Thomas, 2002; Prein, 2002; Primavera 2006), increase food and income security (Edwards et al., 1988; Gupta et al., 1996; Edwards, 1998; Prein and Ahmed, 2000; Devendra and Thomas, 2002; Tipraqsa et al., 2007; Kremen and Miles, 2012), reduce vulnerability to the effects of climate change, and protect biodiversity (Gurung, 2012).

1.5 Research Methods

1.5.1 Study Area

The data were collected from indigenous households in the plains of northern and northwestern Bangladesh (Figure 1.2). The study area included three districts (Dinajpur, Rangpur, and Jaypurhat) in the north and two districts (Netrokona and Sherpur) in the northwest. The study sites included ten sub-districts/*upazilas/thanas* (Pirganj, Mithapukur, Panchbibi, Birampur, Birganj, Hakimpur, Kaharole, Fulbari, Parbatipur, Nawabganj) in the northern districts and four sub-districts/*upazilas/thanas* (Kalmakanda, Durgapur, Jhenaighati, Nalitabari) in the

---

3   Rice or fish monocultures may be subject to alternative risks (Naylor et al., 2000).

northwestern districts. Most of the study area is near the border and rural, where normally the indigenous people of Bangladesh reside.

Figure 1.2: Map of the study area indicating the geopolitical districts (purple) and sub-districts (green) in Bangladesh

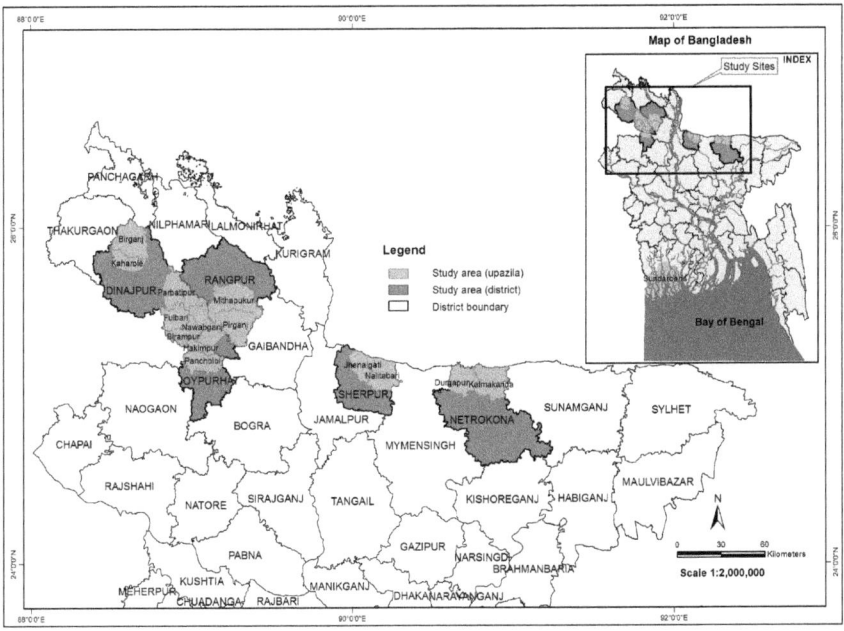

The mean household size among the study sites ranges from three to four members. Literacy rate varies between 35 and 60%. Literacy levels were higher in the northern region relative to the northwest. Households in the northwest own a higher number of ponds than households in the north. The main crop in the study areas is rice (paddy) and agriculture is the main income source for 60% to 70% of the households studied. Thus there is considerable potential for the integration of aquaculture into existing ponds and rice cultivation areas. Mean annual temperatures range from 10°C to 22°C among the study sites. Rainfall is relatively higher in the northwest relative to the northern region. Irrigation coverage varies widely among sites and average cropping intensity ranges from 150% to 210%, some are very close to the national mean of 191%, and others fall above and below the national average (BBS, 2011, 2013). Thus there is potential for increasing irrigation coverage and for integrating aquaculture, and subsequently these efforts would increase cropping intensity in sustainable manner. All of the districts in the study area have large water bodies that are well stocked with fish species such as: ruhi (*Labeo rohita*), mrigel (*Cirrhinus mrigala*), kalbaush (*Labeo calbasu*), katla (*Katla katla*), and indigenous fish species like shoil (*Channa striatus*), shing (*Heteropneustes fossilis*), and koi (*Anabas testudineus*) (BBS, 2013; Banglapedia, 2014).

1.5.2 Data: Sampling Technique, Sample Size, and Survey Effort

The study used household-level three-wave panel data and cross-sectional survey data from the study area (Figure 1.2). The first and second survey rounds were conducted in 2007 and 2009 under the supervision of WorldFish (WF), Bangladesh researchers. The third survey round (re-visited) was conducted from July 2011 to January 2012 by the author himself with the assistance of trained enumerators. The study sites were deliberately sampled from the *Adivasi* Fisheries Project (AFP) sites.[4] A multistage sampling procedure was applied for the survey effort. At the beginning of the AFP in 2007, WF conducted a census of 5337 *Adivasi* households across five districts in northern and northwestern Bangladesh. Out of the total sample, 3594 extremely poor households (based on wealth ranking) were selected from 120 communities for intervention by the project. A total of seven suitable livelihood intervention options within IAA value chains were disseminated among the selected households according to their resource base, and social and economic characteristics such as income, land holding size, and food security status. Among the selected households, those that were relatively wealthier and that own or have access to suitable assets for fish culture (i.e. ponds, rice fields, community aquatic resources) were engaged in IAA production related value chain interventions. The relatively poorer households, such as those that were landless or that lacked significant physical or economic resources, were selected for inclusion within upstream and downstream IAA value chain activities such as fingerling or fish traders and fishermen (netting). Entire sample households received technical training through a 'farmer field school' (FFS) and initial financial support (AFP, 2010; Pant et al., 2014).

To assess the nature and extent of changes resulting from IAA value chain participation WF, Bangladesh conducted a random survey with 510 of the participating households and 147 non-participating households (control) in 2007 (first round of survey). That survey effort employed a structured questionnaire

---

4 The project was implemented by WorldFish and its partner organisations from 2007 until 2009 to increase food security and dietary nutrition by diversifying livelihood options among resource-poor, marginalized *adivasi* (indigenous) communities (see Pant et al., 2014).

featuring information about the asset base and livelihood portfolios of the households surveyed. WF conducted a follow-up survey (the second round of survey) with the same households in 2009 using the same questionnaire to monitor impacts at the household level. The author revisited the same households in 2012 (the third round) and surveyed a subset of 450 participating households and 121 non-participating households. Table 1.1 describes the sample and dynamics over time. Based on data from the third survey round it is evident that the IAA value chain participating households split-up into two groups, those that continued participation in IAA value chain activities (234 households) and those that had abandoned participation in IAA value chain activities (216 households). This reflects the dynamics of IAA value chain participation. In this study we explored the factors that determine participation dynamics and welfare impacts. This appears to be the first analysis of long-term panel data on IAA systems that considers all value chain actors. Between the first and second survey rounds there was no sample attrition, but between second and third rounds there was some sample attrition, which is normal for long-term panel surveys. The sample attrition in the third survey round was 13.1%. We tested for attrition bias and found that it was random (see the attrition bias test results in the appendix in Table A.1.1). Due to migration, death, and regular absence from the home, some sample households from the first and second survey rounds could not be included in the third round survey. Throughout the dissertation we treat IAA participation as 'technology adoption,' and in accordance with the treatment of technology adoption categories in the literature (as either 'adopters,' 'dis-adopters,' or 'non-adopters') we describe members of IAA participation categories as 'participators,' 'dis-participators,' or 'non-participators.'

Table 1.1: *Sample size of the panel survey of integrated aquaculture-agriculture participation among study area households in Bangladesh*

Survey round	Year	IAA value chain non-participators	IAA value chain participators	IAA value chain dis-participators	Total	Attrition (%)
1st Wave	2007	147	510	–	657	–
2nd Wave	2009	148	509	–	657	–
3rd Wave	2012–2013	121	234	216	571	13.09

For the third survey round recent agricultural and fisheries graduates were hired and trained as survey enumerators, and interviews were conducted in the local languages under the supervision of the author. In addition to the questions asked in the first two survey rounds, we included detailed questions on production, revenues, and the costs of IAA value chain activities. In the third survey effort we also included questions on perceptions of the social and environmental impacts of rice–fish based IAA and rice monoculture systems. The sample for the third survey round included an additional 133 non-indigenous rice monoculture farmers (see the third survey questionnaires Table A.1.2 in the Appendix). The analyses of the data derived from the additional questions and sample households are discussed in chapters 2 and 5.

1.5.3 Indigenous People of Bangladesh

Bangladesh occupies the Ganges River delta and is one of the poorest and most densely populated (1203 ind./km^2) countries in the world. The amount of arable land per capita is only 0.05 ha with a population of over 156 million and an area of 147570 km^2 (Badiuzzaman et al., 2013; BER, 2014; WDI, 2014). Of the total population, around 2% (more than two million) are indigenous people living in the border areas of the Northwest and Northeast Chittagong Hill Tracts (CHT). There are more than 49 indigenous communities with distinct cultural, ethnic, religious, and linguistic identities. Broadly, they can be classified as belonging to two groups, one inhabits the plains of the northern and northeastern Bangladesh and the other is the *'Pahari'* or *'Jumma'* (hill tribes) concentrated in the CHT (Barkat et al., 2009; Roy, 2012). The former indigenous group is featured in this study.

Although there is a commonly accepted definition of indigenous people, in Asia and particularly Bangladesh prespective, the Asian Development Bank (ADB) working definition of indigenous people seems more appropriate. According to the ADB working definition, indigenous peoples are those that display two significant characteristics: "(i) descent from population groups present in a given area, most often before modern states or territories were created and before modern borders were defined; and (ii) maintenance of cultural and social identities; and social, economic, cultural, and political institutions separate from mainstream or dominant societies and cultures" (Plant, 2002; 7). In Bangladesh the indigenous people originally lived in sparsely populated areas and had ready access to natural resources, mostly in the border areas. Like many other countries, in Bangladesh indigenous people have been historically subjugated and discriminated against, and are the most marginalized social group in the country (Pant et al., 2014).

The socio-economic status of Bangladesh's indigenous is typically marginalized and very poor, with a higher frequency of health problems, nutritionally poor diets, and poor hygiene relative to the rest of the country. In general the socio-economic status of the plains indigenous group is relatively worse than that of the CHT indigenous group (Roy, 2012). Indigenous people in Asia and many other parts of the world are often geographically, politically, socio-culturally, and psychologically marginalized (PRSP, 2008; Tauli-Corpuz and Malanes, 2010). Originally the indigenous groups engaged in diversified livelihood strategies that included crop and livestock farming, fishing (and the harvest of other aquatic animals such as crustaceans and molluscs) in wetlands, and hunting in forests (of small terrestrial animals and birds), but in recent times traditional livelihood strategies are under threat due to a combination of social, economic, and ecological factors, such as land grabbing, declines in the productivity of natural fisheries and forests, socio-economic marginalization, and exclusion from development programmes and policies (e.g. social safety net programmes). WF and its partners implemented the AFP in the north and northwest of Bangladesh ('plains' indigenous group areas) during 2007–2009 in order to diversify livelihood options, reduce the vulnerability of households, and increase resilience through the development of IAA value chain activities in indigenous communities (Pant et al., 2014).

1.6 Outline of the Dissertation

The introductory chapter presents the overall problem statement, research questions, conceptual framework, study area, survey data, and an overview of the indigenous people in Bangladesh. The history and characteristic features of the indigenous people of Bangladesh and how they differ from the rest of society in terms of social, ecological, economic, and cultural aspects and the general socio-economic status in Bangladesh are also described in this chapter. The chapter also includes a broad discussion of the data and study sites to give an overview of the data collection and planning process of the entire study.

Rice–fish based IAA value chains are addressed explicitly in Chapter 2 along with a discussion about the financial performance of different value chain actors, the value addition process and function, and gender disaggregated employment along value chains. The chapter includes a discussion of the factors that contribute to success or impose barriers with respect to IAA value chain development within a broader discourse on sustainable agricultural development. In Chapter 3 we explore the value chain participation dynamics in greater depth. In this chapter we identify the determinants that distinguish among IAA value chain participators, non-participators, and dis-participators (those who discontinue participation over time), the latter of which is very often overlooked in technology adoption research.

Chapter 4 features analyses on the welfare impacts of IAA value chain participation. This includes a description of the linkages between IAA value chain participation and the welfare of indigenous households. The discussion also highlights the distributional effects of IAA value chain participation based on impacts on all value chain actors.

Chapter 5 presents the evaluation of rice–fish based IAA socio-environmental impacts and the differences relative to rice monoculture systems that have dominated rice production and received significant policy support in Bangladesh since independence. The analyses presented in this chapter examined the competitiveness of rice–fish IAA relative to rice monoculture systems by considering environmental impacts. Throughout the dissertation emphasis is placed on the application of quantitative research methods in the analyses of data generated by structured surveys to better understand the casual effects of IAA in Bangladesh. The conclusions presented in Chapter 6 summarize the major research findings, attempt to formulate a new research agenda for sustainable development through IAA value chain development in light of these research findings, and identify potentially effective policy options.

Chapter 2: Performance of Integrated Aquaculture-Agriculture Value Chain Development in Bangladesh

2.1 Introduction

With more than 156 million inhabitants and an area of 147570 km^2 Bangladesh is one of the most densely populated countries in the world (with approximately 1203 people per square kilometre or 0.05 ha per capita), rapid population growth (1.37% per annum), and low per capita income (US$1010 per annum) (BER, 2012; WDI, 2014). Although the country's poverty level declined at an impressive rate over the last decade, the absolute number of people living below the national poverty line (US$1.13 per capita per day) remains significant. Around 53 million people still live below the poverty line in the country, most of which (about 75%) live in rural areas (BER, 2012; World Bank, 2012). Agriculture and fishing are major contributors to the economy, accounting for 20% of GDP and have grown by 5% since 2002–2003. Most rural Bangladeshis engage either directly or indirectly in agriculture for their daily livelihoods, and about 48% of the labour force is employed in this sector (BER, 2012). Since ancient times agriculture and fishing have been integral to the lives of the Bangladeshi people and play major roles in food security, employment, dietary nutrition, foreign exchange earnings, and other socioeconomic aspects. Fish with rice is the basis of the national diet, giving rise to the proverb *"Mache bhate Bangali"* (a Bengali is made of fish and rice): fish alone supplies about 60% of the animal protein intake and rice alone supplies 70% of direct daily calorie intake in the national diet (Alam and Thomson, 2001; Sarder, 2007; DOF, 2010). Bangladesh is a major producer and consumer of both rice and fish, which are associated with daily sustenance in Bangladeshi culture, especially among poor rural people.

Due to high population growth, economic development, and urbanisation, demand for rice and fish is constantly increasing. The supply of these staples is threatened due to the conversion of agricultural land (e.g. to roads and houses), the effects of climate change, and negative environmental impacts of fertiliser and pesticide overuse stemming from the GR. Thus there is an urgent need for more sustainable rice and fish production options. Integrated rice–fish farming systems (IRFFS) seem to be such an option, capable of producing more rice and fish in a more sustainable manner by requiring less land and water resources. Since its inception different research efforts have shown that relative to conventional agriculture IRFFS can: be environmentally friendly; function as an integrated pest management (IPM) practice; increase soil fertility; make more optimal and complementary use of scarce land and water resources; increase productivity; be more sustainable; contribute to agricultural biodiversity, intensification, and diversification; and can improve household dietary nutrition (Coche, 1967; Lightfoot et al., 1992; Fernando, 1993; Halwart et al., 1996; Little et al., 1996; Rothuis et al., 1998a, 1998b, 1999; Berg, 2001, 2002; Halwart and Gupta, 2004; Frei and Becker, 2005a; Giap et al., 2005; Gurung and Wagle, 2005; Dugan et al., 2006; Nhan et al., 2007; Haque et al., 2010; Ahmed and Garnett, 2011; Ahmed et al., 2011). Although the potential for IRFFS development in Bangladesh is widely documented, adoption of these farming systems is not widespread (Nabi, 2008; Ahmed and Garnett, 2011; Ahmed et al., 2011). This situation heightens the motivation for proper assessment of the potential socio-economic benefits of IRFFS relative to traditional rice monoculture, as well to identify the factors that facilitate and hinder broader adoption and diffusion.

For the development of effective policy for promoting IRFFS it is necessary to better understand the pattern of its adoption as well as its impacts (Becerril and Abdulai, 2010; Noltze et al., 2012). Doss (2006; 208) mentioned that; "technology adoption research has made substantial progress in the ability to examine the intensity of adoption (not simply based on dichotomous choices) and addresses the simultaneity in adoption of different components of a technology package. The issues of how institutional and policy environments affect the adoption of new technologies and how the dynamic patterns of adoption affects the distribution of wealth and income however, remain unanswered." There do not appear to be any studies that consider these aspects with respect to IRFFS in Bangladesh. The value-chain analysis combined with partial budgeting and 'strengths, weaknesses, opportunities, threats' (SWOT) analyses provide a useful ex-ante tool for assessing the performance of this farming system by considering the upstream and downstream value chain actors (e.g. Veliu et al., 2009; Christensen et al., 2011; Macfadyen et al., 2012). Value-chain analysis is a strong qualitative as well as quantitative approach that has been widely applied to 'pro-poor' economic development issues. It can be used to assess economic viability and sustainability, and to identify critical issues and impasses for different actors, as well as to generate more effective policies and development strategies (Coles and Mitchell, 2011). Thus this study is an attempt to fill research gaps by applying the value-chain analysis as a framework for assessing the relative performance of IRFFS among indigenous households in Bangladesh. An important contribution of this chapter is the application of the value-chain analysis in combination with partial budgeting and SWOT analyses as ex-ante evaluation framework to assess the performance of IRFFS in marginal, extreme poverty settings, which in turn should contribute to the design and execution of agricultural interventions to reduce extreme poverty and marginality elsewhere in the developing world.

The next sections of this chapter present the research methodology, including the data and analytical research methods employed. This is followed by the results and discussion of the value-chain mapping, gross margin, partial budgeting, and SWOT analyses in detail. The chapter concludes with a discussion of the policy implications of the research findings.

2.2 Methods

2.2.1 Data

Data were collected between August 2012 and January 2013 from 12 local geopolitical sub-divisions or '*upazilas*' of the Dinajpur, Rangpur and Joypurhat districts in northwestern Bangladesh and from four *upazilas* of the Netrokona and Sherpur districts in the north (Figure 2.1). The study sites were chosen because of the EU funded AFP at these sites conducted by the WF programme and its associated partner organisation from 2007 to 2009. Participating farmers received training on IAA value chain activities and initial financial support from the project.[5]

The field survey was performed using two types of structured interview schedules over a period of six months. The interview schedule was prepared based on a literature review, pretesting, and expert consultations. Data were collected from actors within existing IRFFS chains. The details of the sample used in this study are shown in Table 2.1. We acquired a list of project participants from WF and randomly selected the sample from that list. Participating farmers were interviewed at their houses or farms. We reviewed the data collected by the enumerators in the field for errors or ambiguity and then cross-validated the data with the same farmers. During the course of the surveys and cross validation efforts the participant observation (direct observation, passive deception) method was used to triangulate the survey data (Bernard, 2006). In addition to primary data, secondary data were collected whenever necessary from government sources like the Bangladesh Department of Fisheries (DOF), the WF, other relevant Bangladeshi ministries, and related literature.

---

5   See Pant et al. (2014) for detailed discussion of the AFP.

Table 2.1: Sample sizes by rice–fish based integrated aquaculture-agriculture value chain participation category in Bangladesh

Category	Sample size	Percent
Fingerling (immature/starter fish) trader	17	4.02
Rice–fish producer	48	11.35
Rice monoculture producer[6]	311	73.52
Fishermen	19	4.49
Fish trader	28	6.62
Total	423	100.00

2.2.2 Analytical Methods

2.2.2.1 Value-chain analysis

Value-chain analysis has become widely used since the early 1990s as a novel methodological tool for examining system dynamics. There is not an exact and universal definition of the value-chain concept. Definitions vary widely depending on the field and scope of study. According to Kaplinsky and Morris (2001; 4) a value chain "describes the full range of activities which are required to bring a product or service from conception, through the different phases of production (involving a combination of physical transformation and the input of various producer services), delivery to final consumers, and final disposal after use." A value-chain analysis focuses on vertical and horizontal linkages among different actors and the movement of goods or services from producer to consumer along the chain. Value-chain analysis is widely used to examine entire industries and recently for research and policy in the agricultural sector as an analytical tool, even in more complex production network environments (Gereffi, 1994; Kaplinsky, 2000; Sturgeon, 2001; Doland and Humphery, 2004). Value-chain analysis can examine values and value addition within chains, the nature of power relations, and power distributions based on governance of the chain, and potential points of entry or exclusion (especially in the case of smallholder farmers), as well as the distribution of revenues and benefits among actors in a chain (Walters and Lancaster, 2000; Wood, 2001; Doland and Humphery, 2004). In addition, in a value-chain analysis it is possible to integrate gender issues (Barrientos et al., 2003), dietary nutrition (Fan and Pandya-Lorch, 2012), welfare, poverty, inequality, and

---

6   Within the rice monoculture category there were 132 non-indigenous farmers that were sampled randomly from local households, the rest of the rice monoculture farmers and survey participants from other categories were indigenous households.

environmental concerns (Kaplinsky, 2000; Gereffi et al., 2001; Bolwig et al., 2010; Riisgaard et al., 2010; Trifković, 2014).

Many value-chain approaches have evolved over time from various disciplines such as economics, environmental studies, and political science (Fasse et al., 2009). Broadly it can be divided into two groups: one that is more descriptive and qualitative that was emphasized by Kaplinsky and Morris (2001), and another that refers to specialised tools and is more quantitative (relies on modelling and simulation), especially in business administration such as efforts to optimize chain logistics (Ontersteijn et al., 2006). Blending qualitative and quantitative methods in a value-chain analysis can include combinations of surveys, focus group interviews, participatory rapid appraisals, informal interviews, and secondary data. In value-chain analyses it is also important to examine institutions, their arrangements, and how they are embedded in the chain to better understand the economic, social, and political implications. In value-chain analysis certain norms, working rules, and proprietary relationships can have significant influence on individual choice, in this case of the particular internal or external stakeholder in the chain. Individual actors must decide whether or not they are willing to participate in the chain. Here, the term governance comes into consideration, which in this case means the transformation of institutions driven by the actors. In this regard governance (systems) shows whether institutions are effective or not (Hagedorn, 2008). As goods and services move along the chain from one actor to another there are (transaction) costs as each good or service is transferred, which might be fixed or variable. This can lead to coordination problems. According to Williamson (1985) there are three determinants of transactions: asset specificity, uncertainty, and frequency. Asset specificity is related to the specific investment in the transaction and how costly the investment is relative to alternative uses of the good or service. The more difficult it is to reallocate the resource to another use, the more specific the transaction is. Uncertainty refers to the uncertain action or behaviour of the contract partner. Frequency indicates the repetition and number of transactions. The more frequent a transaction is, the more trust there is between the relevant actors and the less opportunistic the behaviour.

Various types of analysis can be undertaken using the value-chain approach, such as: functional analysis (Guptill and Wilkins, 2002; Bahr et al., 2004), institutional analysis (FAO, 2005a), social network analysis (Kim and Shin, 2002), financial analysis (FAO, 2005b), input-output analysis (Hecht, 2007), social accounting matrix (Adelmaan et al., 1988; Courtney et al., 2007), life-cycle analysis (Rebitzer et al., 2004), input-output life-cycle analysis (Lenzen, 2001), material flow analysis (Finnveden and Moberg, 2005), energy analysis (Finnveden and Moberg, 2005), and integrated ecological-economic modelling (Baecke et al., 2002; Pacini et al.,

2004; Kledal, 2006). One method cannot cover all aspects of a chain, so in this study we used a combination of methods including functional analysis, which depicts the interactions among actors in the value chain and describes their activities along the chain. We also used institutional and social network analyses to provide an overview of the various chain actors and the relationships among individuals, groups, and organisations in value creation. We used financial and input-output analyses to determine the financial costs and benefits to individual agents along the chain and to trace the flow of goods and services between actors. We also used a material flow analysis to assess the physical units of input and output involved in production, processing, consumption, and distribution.

2.2.2.2 Gross margin analysis

Gross return is calculated by multiplying the total volume of output by the mean price during the harvest period (Dillon and Hardaker, 1989) as follows:

$$GR_i = \sum_{i=1}^{n} Q_i P_i \tag{1}$$

where,
  GR_i = gross return from the ith product (Tk/ha)
  Q_i = the quantity of the ith product (kg)
  P_i = the mean price of the ith product (Tk/kg)
  i = 1, 2, 3..................., n

In farming the financial performance of an activity is usually expressed in terms of gross margin, defined as the difference between gross return and total variable costs. Fixed costs are not included (Nix, 2000). Gross margin can be expressed as:

$$GM = GR - TVC \tag{2}$$

where,
  GM = Gross margin
  GR = Gross return
  TVC = Total variable cost

The benefit-cost ratio (BCR) is a relative measure that is used to compare benefits per unit cost. The BCR was estimated as gross returns divided by total variable costs using:

$$BCR \text{ (undiscounted)} = \frac{\text{Gross return (GR)}}{\text{Total variable cost (TVC)}} \tag{3}$$

2.2.2.3 Partial budgeting analysis

Partial budgeting is normally used to re-evaluate the economic viability of an activity when there is a minor change in a production technique resulting in a partial change in cost-return structure (Barnard and Nix, 1979; Shang, 1986). Partial budget analysis assesses the incremental technological change at the field level (Roth and Hyde, 2002; Holland, 2007). It only includes the resources that will be changed, leaving out those that are unchanged (e.g. fixed assets), and supports the assessment of alternatives. Partial budget is a balance that measures the positive and negative effects of a change in the existing activities (Kay et al., 2008). It shows how adopting a new technology affects profitability by comparing the existing one with new or alternative methods. It is based on the concept that technological change will have one or more positive or negative economic effects. On the positive side it is assumed that the adoption of technological innovation will eliminate or reduce some costs and/or increase returns. On the negative side it is assumed that technological change will cause some additional costs and/or reduce some returns. The net effect of the introduction of technological innovation is measured by the net change between positive and negative economic effects. A net positive or negative change indicates a potential increase or decrease in income/profit respectively due to the introduction of the new technology (William et al., 2012). We used the partial budgeting technique to re-evaluate the economic viability of IRFFS relative to rice monoculture systems.

2.2.2.4 SWOT analysis

As a framework SWOT analysis is widely used due to its simplicity and practicality. SWOT analysis is used to analyse internal and external factors in a systematic approach to provide support for decision making. It is a valuable tool for addressing some of the weaknesses of quantitative analyses. The objective of SWOT is to determine how to maximise the future position of an organisation/business/enterprise/activity such as IRFFS in Bangladesh (Kurttila et al., 2000).

SWOT analysis is a strategic planning tool consists of two parts (FAO, 2006):

1. An analysis of the internal condition (strengths and weaknesses) of an actor. This part only evaluates actual strengths and weakness rather than potential or expected strengths and weaknesses.
2. An analysis of the external condition (opportunities and threats) of an actor. This includes the context (i.e. existing threats) as well as unexploited opportunities and probable trends.

New technologies or production systems like IRFFS are promoted as a potential means of improving the economic, environmental, and health conditions in developing countries. The actual adoption rates of new production systems and technologies, however, are often less than expected and are not uniform (Feder et al., 1985). We used a SWOT analysis to identify and evaluate the constraints and facilitating factors of the adoption and diffusion of IRFFS in Bangladesh in general, and specifically among indigenous communities.

2.3 Results and Discussion

2.3.1 Value Chain Mapping

Value-chain analysis can be used to reveal how linkages between the production, distribution, and consumption of products are interconnected along value chains that represent a network of activities and actors (Kaplinsky, 2000; Sturgeon, 2001; Barrientos et al., 2003). The value-chain approach identifies the "input-output structure or value-added sequence in the production and consumption of a product; the dispersion of production and marketing; the governance structure or power relations that determine how financial, material, and human resources are distributed within the chain; and the institutional framework that identifies how local, national, and international contexts influence activities within chains" (Barrientos et al., 2003; 1512). Governance structures determine how the benefits of participation are distributed along the chain (Gibbon, 2000; Gereffi et al., 2001; Humphrey and Schmitz, 2001). Governance structures can be producer driven and/or buyer driven (Gereffi, 1995). These structures are helpful for identifying how power is exercised within chains (Barrientos et al., 2003).

The rice–fish value chain maps shown in Figure 2.1 provide a schematic representation of the key actors and the product and information flows at given points in time. The horizontal product flows indicate the alternative supply channels, while each vertical level in the value chain describes the productive function. Value chains encompass a network of supply or marketing channels. The chain of actors through which the transaction of goods occurs between producers and consumers constitutes a supply or marketing channel. In other words, marketing channel refers to a pathway composed of various marketing intermediaries that perform such functions as needed to ensure smooth and sequential flow of goods and services from producers to consumers. Marketing channels are alternative routes of product flows from producers to consumers within the chain (Kohls and Uhl, 2002). In Bangladesh fish produced by IRFFS moved from producers to intermediaries to consumers through such channels (i.e. through market intermediaries such as fish wholesalers and retailers). Fish produced from IRFFS needed to move a short distance from the point of production to consumers over a brief period due to its perishable nature, small-scale production, and high local demand. Within the value chain, marketing channels through which the IRFFS produced fish moved are depicted in Figure 2.1. Here we only discuss the fish component of the rice–fish value chain rather than both rice and fish. We assumed that IRFFS brought extra financial benefits to rice monoculture producers. For details of the value-chain analysis of rice see Minten et al. (2011, 2013) and Reardon et al. (2012).

Figure 2.1: Schematic representation of rice–fish value chains in Bangladesh

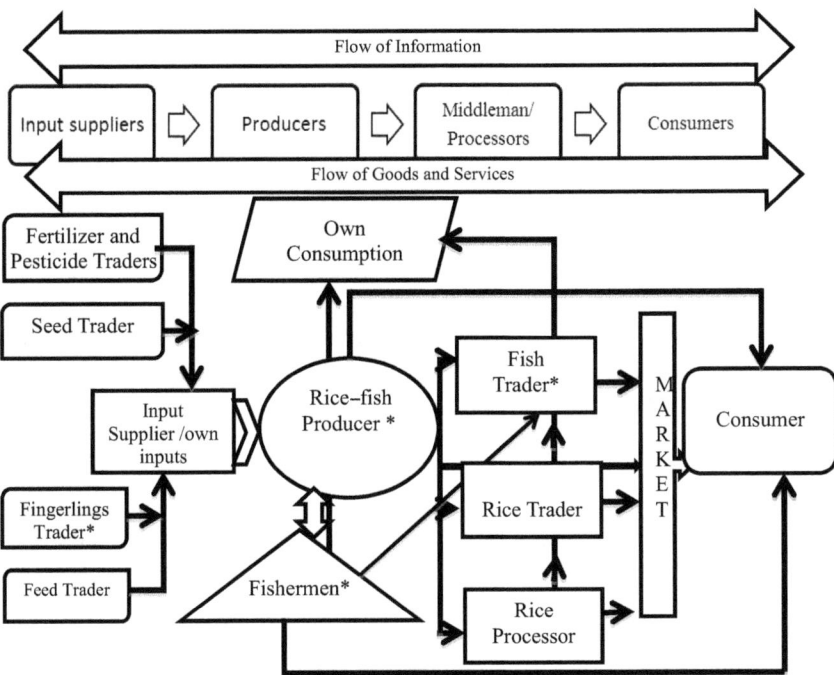

Only a few channels for rice–fish production and distribution were observed and these were very short. Rice–fish farmers use their own inputs or else purchase them from suppliers, such as purchasing fingerlings (immature/starter fish) from fingerling traders and they also use the services of fishermen. Fish produced by IRFFSs are either consumed by producer households and or sold to neighbours, at local markets, or to fish traders. Among these channels the major actors are fingerling traders, rice–fish producers, fishermen, fish traders, and consumers.

All of these actors play important roles along the chain. Fingerling traders are the primary input suppliers to the small-scale rice–fish farmers through credit or cash. In the study area farmers reported that they only bought fingerlings from fingerling traders and that feed was mostly sourced on site although some is purchased from local feed traders or markets. Fish traders purchase fish from rice–fish farms and transport it to local markets. Rice–fish producers have several marketing options for either marketing or consuming the fish that they produce. Producers may sell directly in local markets, to neighbours, or to fish traders. Similarly, fish consumers have the opportunity to purchase fish from different sources.

2.3.2 Actors, Value Addition, Governance, Institutional Framework and Employment in Rice–fish based Integrated Aquaculture-agriculture Value Chains

Major rice–fish value chain actor include input suppliers (fertiliser and pesticide traders, seed traders, feed traders, and fingerling traders), rice–fish producers, fishermen, rice and fish traders, rice processors or millers, and consumers. Normally multiple functions or services are offered by individual actors along the chain such as: exchange functions (buying and selling), physical functions (transport, storage, processing), and facilitating functions (standardization, financing, risk-bearing, and market intelligence) (Kohls and Uhl, 2002). The fish component of the rice–fish value chain is not extensive or complex. Almost all farmed and wild harvested fish are sold either live, fresh on ice, or fresh without ice; there is almost no primary or secondary processing involved. For fish production in rice fields farmers primarily rely on naturally available food sources (e.g. phytoplankton, zooplankton, periphyton, and benthos). Some farmers use additional supplementary homemade feed or fertilisers (e.g. cattle manure, waste rice, rice and wheat bran, etc.,). Some farmers, especially relatively large-scale producers, sometimes purchase feed (e.g. mustard oilcake, poultry manure, fishmeal, commercial pelleted feeds, etc.) from feed traders in local markets. The most common fish species cultivated using IRFFSs are Indian major carp and exotic species such as common carp (*Cyprinus carpio*), katla (*Katla katla*), rohu (*Labeo rohita*), mrigal (*Cirrhinus mrigala*), bata (*Labeo bata*), Nile tilapia (*Oreochromis niloticus*), silver carp (*Barbodes gonionotus*); and various indigenous species. Normally rice–fish producers use relatively large fingerlings. Producers reported that larger fingerlings reach marketable sizes faster than smaller fingerlings. In general, IRFFS producers in Bangladesh do not stock any specific ratio of the different fish species available (Ahmed and Garnett, 2011). Almost all of the actors involved in the production and distribution channels consumed fish throughout the production and distribution periods. Fish is harvested by either fishermen or the farmers themselves and sold to neighbours, wholesalers, or retailers. Farmers sometimes grade fish by species or size to obtain higher or differentiated prices. Most of the time indigenous fish species by-pass established markets and are sold directly to neighbours or else are consumed by the farmers themselves due to local taste preferences and greater demand for smaller and cheaper fish because of limited local purchasing power. Different studies have found that indigenous fish species are typically very nutritious and have the potential to improve food security and dietary nutrition in the developing world (Thilsted et al., 1997; Roos et al., 2003; Roos et al., 2007a; Roos et al., 2007b). Fish traders (wholesalers or retailers) either collect fish from the farmers or have fish delivered to them and then sell fish to retailers, consumers, and restaurants.

Once fish has been harvested from the farm there are no distinct channels for different species (i.e. fish traders deal in all fish species). All rice–fish farmers reported producing and selling a mix of fish species, but carp species predominated. Almost all fish is sold live, although sometimes it is sold fresh on ice (especially during summer months or if the buyer is relatively far from the farm). There is a growing preference among consumers for live and fresh fish, particularly for wild fish relative to farmed fish. Thus IRFFSs have the potential to fulfil these demands in Bangladesh and other developing countries. Actors involved in buying, selling, transportation, processing/grading, cooling/icing, pricing, etc., add value to the product by fulfilling those functions and receive a portion of the marketing margin (see next section for details). Their value addition functions are limited due to the very short spatial and temporal dimensions of the post-harvest portion of the chains.

Fish produced under IRFFS value chains is governed by the spot market transactions involving a large number of small traders, which is the case in many traditional agricultural commodity value chains in developing countries. Modern value chain governance, however, is based on the use of standards and safety protocols throughout chains and high levels of vertical coordination (such as contract farming) and coordination among suppliers and processors (Maertens and Swinnen, 2010). Power relations among the actors are almost equitable because each actors operational units are small. The price for fish is determined through bargaining among actors and bargaining power depends on supply and demand, as well the number of buyer and sellers. Fish traders have more bargaining power when they sell directly to consumers, but when they purchase from producers bargaining power is almost equitable because the numbers of fish traders and producers are limited due to the small scale of the production and marketing.

Information flow within rice–fish value chains is quite transparent. All of the actors exchange information, typically via mobile phones (owned or rented). Although there are several legal restrictions and law enforcement agencies that monitor the quality of food products in Bangladesh, the commercial fish available in urban areas is often adulterated (typically by traders) using poisonous chemical preservatives (e.g. formalin) so that the fish appear to be fresh (Uddin et al., 2011; Rahman, 2013). Fish produced via typical commercial aquacultural systems have relatively longer value chains, making this phenomenon more of a problem. Thus large-scale adoption and diffusion of IRFFS could help reduce the practice of chemical fish adulteration and associated health risks in Bangladesh. Fish farmers and traders in the study area reported that they were not aware of any food standard issues.

There are several, mainly governmental, institutions in Bangladesh that interact with rice–fish value chains. The Ministry of Agriculture (MOA) supports farmers by providing extension services and technical support to agricultural producers other than aquaculture and fisheries actors, and the Ministry of Fisheries and Livestock provides more or less the same services to both the fisheries and livestock industries. The DOF is specifically responsible for fisheries industry activities (i.e. extension services, quality, inspection standards, etc.). There are several national organisations like the Bangladesh Rice Research Institute (BRRI), the Bangladesh Fisheries Research Institute (BFRI), and the Bangladeshi programmes of international research organisations (i.e. WF and several NGOs such as CARE) also help to develop and disseminate rice and fisheries related practices and technologies. Most actors reported that they perform their activity individually rather than within an association or organized group (e.g. associations or cooperatives). There are some community-based organisations (CBOs) among the indigenous people, but most of them deal with socio-cultural problems, which are sometimes indirectly linked with rice–fish value chain activities. Rice–fish value chain actors expressed dissatisfaction with the services provided by most of the governmental organisations. So there is plenty of opportunity for strengthening government organisations that will facilitate the adoption and diffusion of IRFFS in Bangladesh.

Table 2.2: Labour allocation patterns (person days/hectare/year) for rice monoculture and integrated rice-fish production systems in Bangladesh

Activities	Gender pattern	Rice-fish Hired Labour	Rice-fish Family Labour	Rice monoculture Hired Labour	Rice monoculture Family Labour	Total Labour Rice-fish	Total Labour Rice monoculture	Difference
Land preparation	Male	6.71	36.97	2.65	8.07	47.37	11.05	-36.32***
	Female	0.00	3.69	0.02	0.32			
Crop establishment	Male	18.45	47.86	19.83	26.52	75.00	53.21	-21.79***
	Female	4.65	4.04	1.80	5.07			
Fertiliser application	Male	1.13	7.52	1.72	8.03	8.99	10.22	1.24
	Female	0.00	0.33	0.03	0.44			
Weeding	Male	6.66	17.46	13.72	20.85	28.89	40.86	11.97**
	Female	0.73	4.04	1.09	5.20			
Pesticide application	Male	0.24	1.91	0.74	4.80	2.15	5.70	3.56***
	Female	0.00	0.00	0.00	0.16			
Feeding	Male	0.00	4.57	0.00	0.00	14.91	0.00	-14.91***
	Female	0.00	10.34	0.00	0.00			
Harvesting	Male	12.62	37.57	22.75	22.78	63.13	52.67	-10.46**
	Female	3.32	9.62	1.50	5.64			
Threshing, cleaning, and processing	Male	4.72	18.47	7.78	15.29	32.04	29.08	-2.95
	Female	1.87	6.98	0.94	5.07			
Selling and purchasing inputs and outputs	Male	0.91	14.14	0.64	5.72	15.09	6.66	-8.44***
	Female	0.00	0.05	0.00	0.30			
Totals	Male	51.45	186.47	69.84	112.06	237.9	181.90	-56.02***
	Female	10.57	39.09	5.37	22.20	49.65	27.57	-22.09***
N		48		311		48	311	

Table 2.2 demonstrates the labour differences between IRFFS and rice monoculture systems. In a traditional double or triple subsistence rice monoculture system, total labour requirements are approximately 209.46 person-days per hectare,[7] which is significantly lower than IRFFS at 287.57 person-days per hectare. Interestingly there are significant differences in labour requirements between rice monoculture and IRFFS for similar operational activities. Land preparation, crop establishment, fish feeding, harvesting, threshing, cleaning, processing, and the purchase and sales of inputs and outputs all require significantly more labour for IRFFS than

---

7   One recent estimate indicated that 234 person-days per hectare are required each year for HYV rice production in Bangladesh (Ahmed et al., 2013).

rice monoculture systems. In contrast fertiliser application, pesticide application, and weeding require more labour under rice monoculture systems. Employment opportunities in different activities under both systems are fairly dominated by men and household labour. Rice–fish systems create a significantly greater amount of employment opportunities for women relative to rice monoculture. Some of the rice–fish farming system operations (i.e. homemade feed preparation, feeding, supervision) are less strenuous labour compared to rice monocultures and are thus a source of employment for women, especially for female household members (Dey et al., 2013).

The value chain analysis findings also indicate that there are very limited post-harvest losses (in contrast to many agricultural value chains, especially in developing countries). Thus the IRFFS fish value chain is a relatively efficient and offers more opportunity for women.

2.3.3 Gross Margin Analysis of Value Chain Actors

Following a system-level approach for an entire agricultural year, the detail costs and returns from IRFFS and rice monoculture are presented in Table 2.3. Rice monoculture farmers require inputs such as human labour, seed or seedlings, ploughing, manure and chemical fertilisers, irrigation, pesticides and other variable cost items that vary with the level of production. In addition to these variable cost inputs IRFFS farmers also require feed and fingerlings.

There was a significant difference in labour inputs between the two farming systems. IRFFS required greater labour input due to additional activities to make sites suitable for integrated rice–fish production such as strengthening dikes, excavating refuges,[8] and additional subsequent activities like feeding fish (see Table 2.2). Ploughing and seedling costs did not differ significantly between the two farming systems. IRFFS farmers stock fish fingerlings at a mean density of 92.48 kg/ha. Farmers reported that they prefer relatively larger fish fingerlings because they have higher survival rates and reach market size sooner than smaller fingerlings. Interestingly there was a significant difference in fertilization rates among farming systems. IRFFS farmers used less chemical fertilizers but more manure and inorganic fertilisers, whereas rice monoculture farmers did the opposite. IRFFS farmers may have used less chemical fertiliser because fish in the rice fields contribute to soil fertility. Both farming systems use liquid as well as concentrated pesticides for preventing pest and diseases.

---

8   Refuges are a form of ditches, sumps, or small ponds in low-lying parts of rice fields where fish can survive during periods of water scarcity.

Table 2.3: Mean quantity and cost/return values of inputs and outputs of different farming systems in Bangladesh

Items	Quantity (mean in kg or ML or person days/ha/year)		Total value (mean in BD Taka[9]/ha/year)		Difference in quantity/value	Factor shares of total revenue and TVC# (%)	
	IRFFS	RM	IRFFS	RM		IRFFS	RM
Variable Costs							
i. Family labour[¤]	225.56	134.26	47066.50	26772.12	−91.30***	18(35)	20(31)
ii. Hired labour	62.02	75.21	12940.69	14997.13	13.19	5 (10)	11(17)
Total labour (i + ii)	287.57	209.46	60007.18	41769.24	−78.11***	23 (45)	31(48)
Ploughing	–	–	9913.52	11421.12	1507.60	4 (7)	8 (13)
Seeds/seedlings	–	–	6718.35	6869.81	151.46	3 (5)	5 (8)
Fish fingerlings	92.48	0.00	17382.20	0.00	−92.48***	7 (13)	0 (0)
Livestock manure	5604.60	4810.44	5170.21	3358.98	−794.15	2 (4)	2 (4)
Chemical fertilisers	359.37	495.37	8410.50	11547.48	136.00**	3.25(6)	8.5 (13)
Irrigation	–	–	11639.09	10228.21	−1410.88	5(9)	8 (12)
Feed	2604.60	0.00	11316.63	0.00		4 (8)	0.0 (0)
Concentrated pesticides/insecticides	0.70	3.70	1041.72	2101.77	3.01***	0.4 (1)	2 (2)
Liquid pesticides/insecticides	300.76	672.66			371.89**	0.00	0.00
Miscellaneous (machinery costs for dike preparation, lime applications)	–	–	1815.05	180.85	−1634.20***	0.7 (1)	0.13 (0.2)
A. Total variable cost (TVC)			86347.95	60705.34	−25642.6***	33.4	44.6
B. Total cost (TC)†	–	–	133414.4	87477.47	−45936.98***	51.59	64.33
Outputs							
Rice (in Maunds)	159.82	233.19	92589.1	128371.1	73.38***		
Rice by-products	–	–	7354.79	7613.93	259.13		
Cultured fish	1149.92	0.00	128628.7	0.00			
Indigenous fish	110.76	–	19868.18	0.00			
Others (e.g. vegetables)	–	0.00	10155.10	0.00			
C. Total return from output			258595.9	135985.0	−122610.9***		
D. Gross margin (C-A)			172247.9	75279.7	−96968.29***		
E. Benefit cost ratio (C/B) (Undiscounted)			2.01	1.84			
F. Return per unit of family labour (D/i)			764	561	−203		

Notes: †Total cost (TC) only includes family labour as a fixed cost and all other fixed costs (e.g. land use costs) are assumed to be the same for both farming systems; # the factor share of total variable costs are in parenthesis; ¤ family labour cost was calculated by assuming family labour wage is same as hired labour; one *maund* = 40 kg; * significance at 10%, ** significance at 5%, *** significance at 1%

---

9 The exchange rate in 2012 was US$1 to 79 Bangladeshi Taka (BDT).

Rice monoculture farmers, however, used significantly more liquid and concentrated pesticides compared to IRFFS farmers. IRFFS farmers mostly used liquid pesticides. Farmers reported that there are some pesticides that do not affect fish survival and consequently that they use those pesticides without any awareness of residual effects. IRFFS farmers must also spend money on miscellaneous costs such as land modifications before and after rice and fish harvests, and some farmers produce vegetables on dikes, which incur additional costs from the purchase of required items such as seed, bamboo supports, rope, etc.

According to the survey data rice monoculture systems had higher mean annual rice yields per hectare (233.19 maunds/9327.60 kg) than IRFFS (159.82 maunds/6392.80 kg). The difference in rice yields between the two farming systems was significant, perhaps due to differences in the use of inputs (seeds, fertilisers, and pesticides) and the intensity of management or need for technical skills. Many farmers reported that rice does not grow as well under IRFFS and that fish sometimes destroy rice plants. Halwart and Gupta (2004) found that bottom feeding carp species, especially the common carp, and herbivorous species such as the grass carp can uproot and consume entire rice plants if they are stocked before the rice plants develop adequate root systems or as large fingerlings. Thus fingerling management is crucial for the success of IRFFS efforts, especially with respect to rice productivity (Nabi, 2008). The mean annual cultured and indigenous fish yields reported by IRFFS farmers were 1149.92 kg/ha and 110.76 kg/ha respectively. Some of the IRFFS farmers cultivate vegetables on dikes, which provide an additional income opportunity.

Overall the total returns and gross margins differed significantly between the systems. Although IRFFS rice yields were significantly lower than rice monoculture yields, IRFFS total returns and gross margins were significantly higher. Thus the rice yield disparity is compensated for by the higher financial return from fish under IRFSS. The resultant increase in gross margins for IRFFS results in a mean benefit cost ratio of 2.01, indicating that, holding all other factors constant, each Bangladeshi taka invested will increase the gross margin by more than two. These results indicate that at the farm level IRFFS appears to be an economically viable alternative to rice monoculture.

A closer look at the factor share of total revenues of these farming systems provides further insight into their economic performance. Factor shares of total revenues and total variable costs explain how the benefits shared among production factors, input intensity, and input prices influence the costs and returns of different systems. For IRFFS (Table 2.3) labour, fingerlings, feed, and irrigation were the most costly inputs and had the most benefits shared among them, whereas in rice monoculture systems labour, fertilisers, ploughing, irrigation, and seed were the most costly inputs and accounted for most of the benefits shared among them. Overall the total returns shared among the variable costs are higher for rice monoculture systems than IRFFS. This implies that IRFFS farmers and the fixed factors of this system contribute to higher profit margin shares relative to rice monoculture.

Table 2.4: Mean costs/returns of inputs and outputs of actors along integrated rice-fish farming system fish value chains in Bangladesh

Items	Total value (mean in BD Taka/year)			Factor shares of total revenue (total costs) (%)		
	Fingerling trader	Fishermen	Fish trader	Fingerling trader	Fishermen	Fish trader
Variable cost items						
Total labour+	36263.84 (208.73)	34908.27 (204.50)	35795.68 (214.60)	25.93 (74.16)	40.50 (79.23)	25.14 (66.13)
Food costs	5186.93	2334.99	6273.66	2.94 (6.70)	2.47 (4.96)	3.46 (11.06)
Transport costs	12127.22	9994.05	6938.33	7.95 (16.00)	7.89 (13.30)	4.46 (12.64)
Miscellaneous (e.g. ice, plastic bags, etc.,)	1611.03	1152.63	4170.52	1.68 (3.14)	0.99 (2.51)	3.01 (10.18)
A. Total variable costs (TVC)	18925.20	13481.70	17382.50			
B. Total cost (TC)#	55189.01	48389.94	53178.18	38.50 (100)	51.84 (100)	36.06 (100)
C. Total return from sales	176806.7	94771.93	204471.7			
D. Gross margin (C-A)	157881.53	81290.26	187089.20			
E. Benefit cost ratio (C/B)	3.20	1.96	3.85			
Average number of activity days per year	87.50	82.47	126.11			
Average quantity per day (Kg)	13.00	2.29	20.91			
Return per labour	756	398	872			

Note: + The quantity of labour per year is shown in parentheses; # the total cost includes only labour costs and all other fixed costs are assumed to be the same for all actors.

Table 2.4 presents the costs and returns as well as factor shares for backward and forward linkage actors along the IRFFS fish value chain. As observed from the value chain maps, the IRFFS fish value chain is short and consequently there are not many value added functions performed by the participating actors, thus the list of their costs is not long, primarily labour, transport, and food. The quantity and value of variable cost items were almost the same for all actors, but total returns and gross margins vary among actors. This is due to differences in the mean number of activity days per year, the mean quantity of transactions per day, and the mean difference between purchasing and selling prices. For fishermen this is largely influenced by the mean catch volume or the mean share of fish they receive from fishing as part of a group. All of the actors commented on the seasonality of their activities due to irregular fish supply, water scarcity (drought), and decreased productivity of common resource fisheries like rivers, canals, etc. Thus those actors cannot completely rely on these activities for their livelihoods, which sometimes discourages them from engaging in these activities or motivates them to consider alternative livelihood strategies. Among these three actors, gross margins were highest for fish traders followed by fingerlings traders and fishermen. The gross margins and benefit-cost ratios exhibited the same pattern.

The results of the analysis of factor shares of total returns and total variable costs (Table 2.4) indicate that labour and transport costs are the two most costly inputs for fingerling traders, fishermen, and fish traders. These two inputs also have the greatest share of total returns from the respective business activities. These factor share results suggest that these activities may have significant potential for households with credit constraints and limited market access, and even landless marginalized households because labour costs are the greatest share of total variable costs.

2.3.4 Partial Budgeting

The potential of any technological innovation can be evaluated by its private benefits and costs. A technological innovation is considered to be economically feasible if the benefits from the technology outweigh the costs. Thus to assess the potential of IRFFS relative to the performance of rice monoculture systems we constructed a partial budget using the cost and benefit information derived from the field survey. The results of the analysis are shown in Table 2.5.

It is evident from the partial budget analysis results that IRFFS increases costs as well as benefits, but that the benefits outweigh the cost increases. Thus net changes in farm income due to the adoption of IRFFS were positive and amount to 96968.29 Bangladeshi Taka per hectare annually at the farm level, but considering the other IRFFS fish value chain actors additional benefits would be much greater. Ultimately, the net benefit of IRFFS is primarily the additional income from fish earned by smallholder indigenous farmers without significant income losses or food security threats from forgone rice cultivation.

Table 2.5: *Partial budgeting analysis results: net changes in gross margins are due to the replacement of rice monoculture with integrated rice–fish farming systems in Bangladesh*

Costs	Tk/ha/year	Benefits	Tk/ha/year
1. Cost incurred for integrated rice–fish systems	86347.95	1. Costs saved from forgone rice monoculture	60705.34
2. Forgone revenues from rice monoculture	135985.00	2. Revenues earned from integrated rice–fish systems	258595.90
Net change (++)	96968.29		
Total	319301.20	Total	319301.20

2.3.5 SWOT Analysis of Integrated Rice–fish Value Chains

The SWOT analysis characterised IRFFS based on stakeholder interactions, field observations, in-depth farm household surveys, and a literature review (Table 2.6). The SWOT analysis explored how the IRFFS fish value chain performance could be improved by identifying the critical factors affecting chain performance. All of the issues identified by the analysis are discussed in detail in the subsequent section and represent potential areas of intervention within the value chain and relevant external factors (e.g. policy makers, research organisations, extension agents) to improve value chain performance.

Table 2.6: *SWOT framework results for integrated rice–fish farming system value chain development in Bangladesh.*

Strengths (S)	Weaknesses (W)
Sustainable agricultural development is already on the nation's political agenda	No national strategy has been defined for the implementation of sustainable agricultural development
Multifunctional agricultural system with multiple benefits	Start up costs are high for poor farmers
Can act as an important element of integrated pest management	Requires continuous supervision
	Requires more labour than conventional rice production
Requires less fertilisers, pesticides, and herbicides	Requires greater technical knowledge and skills than conventional rice production
Relatively more efficient and complementary utilisation of land and water resources	A lack of backward and forward linkage actors and their inputs
Increases soil fertility	A lack of broader irrigation coverage
Offers integrated resource management options	Confusion and duplication of

Strengths (S)	Weaknesses (W)
Rice and fish provide a more complete nutritional subsistence benefits to agricultural households	Confusion and duplication of responsibilities among various agencies involved in rice and fisheries management at the central and local levels
Traditional dietary importance (of rice and fish) in rural Bangladeshi livelihoods	A lack of efficient and motivated expertise, resources, budget, and equipment for public agencies
More gender equitable employment opportunities including family labour and lean period employment opportunities	A lack of systems thinking and coordination among crop and fisheries related institutions/policies
Several direct and indirect policies to support integrated rice–fish system improvement in Bangladesh like 5-year plans, the Country Investment Plan, Poverty Reduction Strategy Paper, Protection and Conservation of Fish Act, National Fisheries Policy, National Water Policy, National Agricultural Policy, National Land-Use Policy, and the New Agricultural Extension Policy	Limited or a lack of available component technologies for different rice–fish management options
	A lack of timely availability and quality of fingerlings
	A lack of post-harvest processing and storage facilities
Several agencies involved in crop and fisheries management and officially concerned with developing crop and fisheries technologies, specifically integrated rice–fish systems	Inadequate infrastructure (e.g. roads, markets)
	Requires suitable bio-physical conditions
Availability of competent institutional support such as BFRI (fisheries), BRRI (rice) and WF (fisheries and aquaculture)	Requires greater collaboration among policymakers and development practitioners (related to rice, fish, land, water, and the environment)
Existence of the Department of Agricultural Extension (DEA), National Agricultural Research System and the Bangladesh Agricultural Research Council to help disseminate rice and fisheries technology/practices and to provide a forum for communication among different institutions	

Opportunities (O)	Threats (T)
Provides opportunity to obtain financial and technical assistance from international donors to enhance the capacity of public organisations and related human resources	Risk and uncertainty related to climate variability, and related flooding, drought, etc.
Offers more sustainable future increases of rice and fish yields	Risk of losses due to theft, disease, and fish predators such as snakes and crabs

Opportunities (O)	Threats (T)
Provides productive homemade feed waste use options	Increasing landlessness, and tenant farmers and absentee landlords might exacerbate IRFFS adoption and diffusion
Creates more (women and family labour) employment opportunities	Unfavourable land tenure rights, especially for tenant farmers
Improves food and nutritional security, especially hidden hunger	Low land area per capita in the country
Would make additional use of an extensive agricultural extension system	High production costs
Would make more optimal use of increasingly scarce land and water resources	Increased use of fertilisers, pesticides, insecticides, and herbicides
	Issues surrounding irrigation facilities ownership
Possible to conduct successful communications campaigns regarding public health concerns related to the negative effects of rice monoculture systems and the relative benefits of integrated rice–fish production systems	Involves labour intensive production systems
	A lack of available quality feed and higher feed prices than other livestock options
	Insufficient educational levels among farmers for broader adoption
More sustainable agricultural development path than conventional rice production intensification options	Farmers are unconcerned with long-term environmental benefits of intensification alternatives
Would help to meet future demand for rice and fish consumption	Limited effectiveness of agricultural extension services and available information among farmers
Would be possible to introduce innovation into integrated rice–fish production systems such as improvements to genetic potential and management practices	Limited access to timely credit, high interest rates, and unfavourable loan repayment schedules
Would help to conserve nutritious indigenous fish species	Higher fish mortality due to poor water quality, increased water pollution, greater turbidity, lower water levels, and high water temperatures
Offers potential to increase dietary and crop diversity	
Offers potential to introduce related integrated pest management techniques to rice–fish production systems	Conservative society structure (due to low education and religious customs, especially for women)
Offers potential to introduce agricultural insurance to mitigate risks associated with flood and drought	

Sources: Based on data from personal stakeholder interactions, field observations, farm household survey, and Halwart and Gupta (2004), Frei and Becker (2005a), Nabi (2008), IFPRI (2010), Ahmed and Garnett (2011), Ahmed et al. (2011), and Dey et al. (2013).

2.3.5.1 Strengths

IRFFS is feasible virtually throughout the country's rice production areas without major modifications to traditional production methods. IRFFS is not new, but there is considerable potential for improving performance by introducing innovations. Fingerling production (e.g. Barman and Little, 2006), vegetable production, fruit and tree crop production (on dikes) could be enhanced under efforts to promote IRFFS. Food consumption in Bangladesh, especially among the marginalized poor, is dominated by rice and fish, which can drive IRFFS adoption and diffusion to keep pace with the demand for traditional staple foods. IRFFS is a socio-economic and environmentally friendly and more sustainable agricultural intensification and development option compared to rice monoculture systems. IRFFS offers multiple benefits relative to rice monoculture systems such as: improved public health through controlling rice pests, weeds, and mosquitoes, and reducing the use of chemical fertilisers, pesticides, insecticides, and herbicides (which also helps protect biodiversity); facilitation of nutrient cycling; introduction of the use of fish as an IPM tool; making more intensive and complementary use of land and water resources; improved crop diversification and consequently dietary diversity; and improved soil fertility by increasing nitrogen, phosphorus, and organic matter concentrations. In addition IRFFS creates employment opportunities, especially during the lean season (the part of year when rural employment opportunities are traditionally scarce, typically occurring during the pre-harvest period), and offers more gender equitable employment opportunities relative to rice monoculture systems. IRFFS production areas in Bangladesh are normally situated very close to homesteads, therefore female household members can supervise and perform related agricultural tasks (i.e. feeding fish).

Increasing consumer awareness of the health and quality benefits of fresh fish along with culinary preferences for higher quality food products could provide better returns for IRFFS producers and better quality fish and rice to consumers. In Bangladesh, emphasis is given to sustainable agricultural intensification under the sustainable development paradigm with respect to rice and fish farming through several national policies like the Five Year Plans, the Country Investment Plan, Poverty Reduction Strategy Paper, protection and Conservation of Fish Act, National Fisheries Policy, National Water Policy, National Agricultural Policy, National Land-use Policy, and the New Agricultural Extension Policy can help provide momentum for broader diffusion of IRFFS. The traditional importance of rice and fish in rural livelihoods and culinary culture of the Bangladeshi people, are major drivers of broader adoption and diffusion of IRFFS in the country. The value chain analysis results suggest that IRFFS value chains create employment opportunities for poor households through backward and forward linkages that do not exist in conventional rice monoculture systems. In Bangladesh there are several institutions involved in crop and fisheries management, and that are officially mandated to help develop innovation in crop agriculture and fisheries, specifically IRFFS technology among farmers. In addition there are some specialized competent organisations like BFRI for fisheries, BRRI for rice, WF for fisheries and aquaculture technology development and management, as well as the DEA and DOF which have wide networks throughout Bangladesh for disseminating rice and fisheries innovations and provide extension services to farmers. Furthermore, there is the National Agricultural Research System apex organisation, the Bangladesh Agricultural Research Council, which acts as a forum for discussion among different institutions and help to coordinate and monitor different organisations activities. Thus Bangladesh has a number of institutions related to IRFFS, which indicates institutional framework strength for broader adoption and diffusion of IRFFS throughout potential areas in the country.

2.3.5.2 Weaknesses

There are some weaknesses of IRFFS and at the policy and institutional levels that limit farmers' ability to take full advantage of the above-mentioned strengths. There are certain modifications necessary to make rice fields suitable for fish production, which represent significant costs and in many cases farmers, especially marginalized poor farmers such as many indigenous farmers, ability to invest in new production means is extremely limited. Thus the initial investment costs are the major weakness of IRFFS. IRFFS need continuous supervision to avoid fish losses to theft. Vigilance, feeding and maintenance requirements, land preparation, and harvesting fish increase the labour requirements of IRFFS relative to rice monoculture systems. For broader adoption potential integrated rice–fish farmers need assistance creating suitable bio-physical conditions like improving the water retention capacity of soil and soil quality (soil texture, topography and depth), and working with neighbouring farmers to avoid accidental applications of fertiliser and insecticide on their production areas. Without water IRFFS adoption is not possible, and there is not sufficient irrigation capacity to cover all of the rice producing areas in Bangladesh. Integrated rice–fish fields require more and continuous water than conventional rice monoculture systems. In Bangladesh there is a well-established water market, thus most farmers must depend on water vendor, which is a constraint or weakness that challenges broader adoption of IRFFS. Technical knowledge is very important for broader adoption and diffusion of IRFFS, but providing adequate training to appreciable numbers of uneducated and marginalized poor farmers in Bangladesh is a considerable challenge. Technical knowledge and proficiency required for IRFFS adoption include number of practices such as the modification of fields and farms, proper timing of the stocking of fingerlings into flooded rice fields, and determining appropriate species combinations and quantities of fingerlings.

Depending on the IRFFS characteristics desired, farmers need high quality and timely availability of fingerling stock, but this is quite difficult in many areas, especially during the dry season, and is a significant expense for poor smallholder farmers. Many of the functional roles of backward and forward linkage actors involved in the IRFFS fish value chain are not recognized as legitimate jobs in Bangladeshi society, which discourages people from entering these occupations. This limits the development of fingerling production and supply, fish trading, as well as overall value chains.

Policy and institutions related to IRFFS in Bangladesh also suffer from weaknesses that inhibit broader adoption and diffusion. There are several policies in support of the country's sustainable agricultural development goal, but no strategy has been defined for the implementation of sustainable agricultural development. Like many other developing countries there is a general lack of systems thinking and coordination among related policies and their application in Bangladesh. There are also many organisations involved in rice and fisheries management at central and local levels, but roles and responsibilities of these organisations are not well defined, which creates confusion and redundancy. Among these organisations there is a lack of efficient and motivated expertise, resources, financial capacity, and equipment to effectively promote IRFFS research and dissemination. These organisations also generally lack emphasis on systems thinking and coordination among one another. Historically, investment in the social, economic, and policy dimensions of IRFFS is negligible. As a result, post-harvest processing facilities are not well developed or widely available in Bangladesh.

2.3.5.3 Opportunities

There are good opportunities for IRFFS in rural areas of the country, as most farmers are already engaged in rice production and there are large areas of low-lying rice fields that are suitable for IRFFS. The potential suitable area for IRFS in Bangladesh could be further expanded by improvements to irrigation capacity. Growing awareness among fish consumers about the quality and high demand for live and fresh fish, as well as increasing purchasing power, could provide stimulus for broader IRFFS development. Countries like Bangladesh, which has a very large labour pool, and high levels of unemployment and underemployment, will find IRFFS an attractive agricultural employment option. Moreover there is the problem of a seasonal unemployment that exacerbates seasonal food security issues like *monga*,[10] which can be mitigated by agricultural diversification through IRFFS due to greater variety of production and harvesting schedules that contribute to greater demand for labour inputs. The environmental benefits for both biodiversity and human populations are great advantages of IRFFS that have yet to be properly evaluated.

There are numerous opportunities to obtain financial and technical assistance from international donors to enhance the capacity of public organisations to support broader adoption and diffusion of IRFFS in particular, and for agricultural technology in general, due to the potential of IRFFS as a sustainable intensification option, which is gaining momentum in the international community under the sustainable development goals. Introducing fish production to rice farming creates opportunities for more sustainable use of increasingly scarce land and water resources for simultaneous carbohydrate and animal protein production, which can enhance food security and dietary nutrition in Bangladesh and other potential countries. IRFFS also creates opportunities to increase rice yields more sustainably than through GR approaches to agriculture, which can help the country keep pace with the soaring demand for food, especially rice.[11] Bangladesh is one of 34 countries that face severe dietary nutrition and food security challenges (Ruel and Alderman, 2013). The fish species cultivated in rice fields, especially native species, are rich in micronutrients that can help mitigate chronic hunger problems, especially hidden hunger in Bangladesh and other developing countries. Subsistence IRFFS farmers can make more effective use of a variety of household and small farm waste to produce homemade fish feed (e.g. waste rice, wheat and rice bran, etc.,) to reduce feed costs. The IRFFS value chains create additional channels, actors, and networks, which ultimately create more livelihood and employment opportunities, especially for the marginalized rural poor. Although IRFFS farming is a potential form of agricultural diversification in terms

---

10 Seasonal hunger is locally known as *monga*. At this time of year many households suffer from exacerbated poverty condition, which is very common in Bangladesh and other South Asian and Sub-Saharan African countries (Bryan et al., 2014).
11 The effects of integrated rice–fish systems on rice yields are not yet clear. In the study area rice yields from rice–fish systems were lower than rice monoculture systems. Other studies in Bangladesh and other countries have found positive, negative and neutral effects (see Chapter 5).

of socio-economic profitability, employment generation for women, and food and nutrition security enhancement, the rates of adoption and diffusion are very low.

In Bangladesh the system of extension services provided by the DEA and DOF is quite extensive and covers almost all of the country's sub-districts, so it is possible to take advantage of this extensive system for greater dissemination of IRFFS. By engaging this extension system and other institutions related to rice and fish production it may be possible to conduct a successful communication campaign for increasing public awareness of the health concerns and negative environmental consequences of rice monoculture systems and the relatively positive aspects of IRFFS. It would likely be possible to develop public-private partnerships to more effectively implement IRFFS expansion efforts. Many nutritious native fish species are nearly extinct due to the unintended negative consequences of GR technologies and practices. IRFFS creates opportunities to conserve these fish species, to introduce fish-based IPM practices that could reduce rice production costs, and to offer a more environmentally friendly pest control option. IRFFS increases agricultural diversity by introducing fish species to rice production and the possibility of vegetable or tree production on dikes, which would further increase income and food security. These features make IRFFS a climate change resilient farming system. Although IRFFS is a traditional practice in parts of Asia, including some areas of Bangladesh, it is possible to introduce innovation to this system such as improvements to the genetic properties of rice cultivars and management practices to improve agricultural productivity. IRFFS can enhance institutional innovations like collective management, community-based management of common resources like water, especially in the low-lying areas during the rainy season. As IRFFS is vulnerable to climatic shocks like drought and flood, the broader use of agricultural insurance in Bangladesh to mitigate natural disaster related losses would enhance broader adoption and diffusion.

2.3.5.4 Threats

Obvious threats to IRFFS are the risks and uncertainty associated with increasing climate variability, floods, drought, poaching, and inadvertent poisoning, which are all relatively common phenomena in Bangladesh. Losses to disease and fish predators such as snakes and crabs can cause significant economic damage to producers. Poor water quality, water pollution, turbidity, water scarcity, and high water temperatures also contribute to fish mortality, which ultimately reduces motivation to adopt and continue practicing IRFFS. Reduced access to increasingly scarce natural resources such as water and land are also major threats to IRFFS expansion, especially among poor rural farmers. Landlessness and a land tenure system featuring tenant farmers and absentee landlords are increasing in Bangladesh (Ahmed et al., 2013). Existing land tenure and land property rights, especially land use rights of tenant farmers, are unfavourable for the expansion of IRFFS in Bangladesh.

Due to mounting population growth per capita land area is declining, land fragmentation is increasing, and the increasing costs of farm inputs threaten broader IRFFS adoption and diffusion, especially among poor farm households. To keep pace with the growing demand for food, especially rice, Bangladeshi rice farmers are intensifying rice monoculture production (with up to three crop cycles per year) by increasing the use of fertilizers, pesticides, insecticides, and herbicides, all of which are major threats to IRFFS. Irrigation, especially during the dry season, is mandatory for both rice monoculture and IRFFS, but not all farmers own irrigation facilities or equipment. Most farmers rely on irrigation water markets, which also poses a threat to IRFFS because these systems needs a comparatively greater and more continuous water supplies, which is sometimes very difficult to procure without the necessary irrigation facilities.

Although relatively labour intensive IRFFS creates employment opportunities, this is also a threat because the opportunity costs of labour in Bangladesh are increasing and as an emerging economy the labour force is shifting from agricultural to non-agricultural sectors, which may cause labour crises in agriculture sector in the future. Large-scale commercial IRFFS would require additional sources of commercial feed. As IRFFS and aquaculture are not widespread, the availability of high quality feed and of feed during critical stages of the rice–fish production cycle also threaten broader adoption and diffusion. Feed prices are also very high and increasing, which increases production costs and is ultimately prohibitively burdensome for subsistence farmers. Credit facilities, especially for agricultural purposes, are very weak and interest rates are typically very high. Terms and conditions for credit are not favourable for agricultural intensification because farmers cannot repay loans on a short-term basis because agricultural practices like IRFFS follow seasonal production cycles. IRFFS is intensive in terms of knowledge and technical support, but education levels in general and among farmers in particular, are very low in the country. Limited education is also sometimes linked to farmer awareness of effects, especially environmental effects, of innovative technology or practices. Effective agricultural extension services could fill these gaps, but agricultural extension services in Bangladesh are quite poor despite relatively extensive networks. Governance of agricultural extension systems is also very weak, which ultimately threatens overall adoption and diffusion of agricultural technology, particularly IRFFS.

2.4 Conclusions and Policy Implications

Like many other Asian countries Bangladesh is considered a 'rice–fish' societies because both of these staple foods are part and parcel of Bangladeshi culinary identity, which is highlighted by the popular Bengali saying "*Mache bhate Bangali*," (Rice and fish make a Bengali) (Dey et al., 2013). Estimated suitable land area for IRFFS in Bangladesh range from two to three million hectares (ADB, 2005; Dey and Prein, 2006; Ahmed and Garnett, 2010; Dey et al., 2013), yet currently only approximately 0.18 million hectares under IRFFS (Dey et al., 2013). This raises questions about whether or not the overall performance and potential constraints on the adoption and diffusion of IRFFS into these potential areas have been properly assessed. We attempted to provide insight into IRFFS in Bangladesh using a value chain analysis framework with detailed data from recent surveys of IRFFS fish value chain actors in indigenous settings.

We examined the financial performance of different actors of rice–fish value chains using a gross margin analysis. We further investigated whether the integration of fish into rice production would improve profitability and justify efforts to introduce relevant innovations. We conducted a partial budget analysis for two different rice production systems; a conventional rice monoculture production system and an integrated rice–fish production system. We also explored the internal and external factors to further improve IRFFS and its broader adoption and diffusion. We used a SWOT analysis to evaluate technology, policy and institutional level strengths, weakness, opportunities, and threats associated with IRFFS, which can provide insight for future strategy building about its promotion.

The research findings suggest that IRFFS offers considerable potential for increasing overall agricultural productivity and farm income. IRFFS value chains provide opportunities for poor landless households to participate profitably in related value chain activities. This appears to have been the first attempt to comprehensively examine the costs and returns of IRFFS value chains at a national level. In addition to profitable business opportunities, IRFFS provides employment opportunities for women, especially for rice and fish production. Employment related to IRFFS production is further shaped by the social and institutional context within which they operate. Although IRFFS potentially creates employment opportunities for women, resulting effects on household labour allocation and reproductive roles requires additional attention. The partial budgeting analysis findings support broader findings that IRFFS can be an economically competitive alternative to rice monoculture. In addition to potentially empowering value chain actors, IRFFS also offers both direct and indirect benefits to local communities, and opportunities to enhance rural economic growth.

However, IRFFS faces a number of significant challenges and its adoption and difusion has been sluggish in Bangladesh. Experimentation by some innovative farmers, private initiatives, and support by NGOs have been key drivers of the integration of IRFFS into mainstream agriculture. There is a virtual lack of government support for IRFFS farmers and overall value chain development. The high initial costs of land, labour, fingerlings, feed, and modifications required to integrate IRFFS into existing rice production systems are major constraints to broader adoption and securing related pro-poor benefits. In the short run, non-production IRFFS value chain actors have fewer entry barriers and if combined with IRFFS farming the benefits could be significantly higher over the long run for poor farmers despite the high initial costs. The traditional strengths of the systems along with abundant water, fertile soils, strong research and extension institutions, expanding infrastructure, and supportive government policies are failing to overcome numerous weaknesses and threats. Indeed the enormous opportunities for further improvements to IRFFS technology and related value chain performance provide a strong argument for private sector participation within value chains, and government in the form of supportive policy and legislation (on issues such as land tenure; access to credit, markets, and quality irrigation water and feed; required infrastructure; and public and private sector capacity building). Such actions would serve to both safeguard the current levels of the practice of IRFFS as well derive additional benefits from increased adoption in the future.

Value chain analysis has not been widely used to assess ex-ante performance of integrated farming systems in general, and of integrated rice–fish production in particular to explore the potential for their further development. The research findings show how value chain analysis together with SWOT analysis can help to better understand the financial and social benefits of IRFFS by identifying the constraints that hinder greater adoption and overall value chain development. IRFFS production and its value chain development in developing countries could give momentum to the sustainable agricultural intensification paradigm as these systems have traditional strengths and opportunities for further development, however, constraints should be acknowledged and addressed to make this happen. An important contribution of this study is that it was conducted in an indigenous setting, thus the findings of this study may be helpful for the design of agricultural interventions to reduce extreme poverty and marginality in Bangladesh and other countries with similar socioeconomic, agro-ecological, and institutional conditions.

Chapter 3: Integrated Aquaculture-Agriculture Value Chain Participation Dynamics in Bangladesh

3.1 Introduction

Agriculture continues to be the predominant livelihood option in many of the world's poorest countries. However, agriculture in many countries faces numerous threats, such as low or declining productivity due to land degradation, climate change, and trends of other environmental changes. Given the increasingly limited scope for further horizontal expansion and changing climatic and environmental conditions in many developing countries, there is no substitute for a sustainable agricultural intensification strategy. One important strategy for increasing productivity is through the adoption of improved and more sustainable agricultural technologies and management (Baffes and Gautam, 2001; Doss, 2006; Amsalu and Graaff, 2007; Dey et al., 2010; Kijima et al., 2011). One potential approach to sustainable agricultural intensification is IAA farming systems that have been promoted in many parts of Africa and Asia by different national and international research organisations and under public, private, and NGO initiatives. One notable effort is the WF participatory development and dissemination approach that began in the 1980s and has identified significant potential benefits of IAA adoption (Dey et al., 2010; Dey et al., 2013). Even though IAA seems to have considerable potential for building the momentum of the sustainable intensification paradigm in Africa and Asia, adoption rates are still very low (Nabi, 2008; Ahmed and Garnett, 2011; Ahmed et. al., 2011).

Thus it is crucial to better understand how new technologies are adopted in practice if their potential is to be fulfilled. The adoption and diffusion of innovation are measured by the observed adoption rates and have long been studied in the context of technological change, economic development, poverty reduction, and food security (Feder et al., 1985; Dinar and Yaron, 1992; Besley and Case, 1993; Graaff et al., 2008). Until the mid-1990s most technology adoption studies were based on cross-sectional rather than panel data, based on the consideration of technology adoption as a static process,[12] although long time before the seminal work of Griliches (1957) and Mansfield (1961) described technology

---

12 See Mahejan and Peterson (1985) for further a description of drawbacks of the static adoption and diffusion models.

adoption as a dynamic process (Besley and Case, 1993; Barham et al., 2004; Doss, 2006). Significantly, recent studies have highlighted the advantages of using panel data to study the dynamics of technology adoption (Besley and Case, 1993; Foster and Rosenzweig, 1995; Cameron, 1999; Conley and Udry, 2001; Barham et al., 2004; Doss, 2006).

The purpose of this study was to contribute to the existing literature by examining the processes of participation, non-participation, and abandonment of practicing IAA and its related value chain activities among marginalized indigenous households in Bangladesh. IAA is an integrated resource management practice that is broadly defined as two or more linked farming activities, of which at least one is aquacultural, and in which outputs and inputs from one subsystem that otherwise may have been wasted or underused become an input to another subsystem, resulting greater efficiency (Edwards, 1987; Edwards et al., 1988).[13] The research data span the entire period from the initial introduction of IAA and related enterprises in 2007 until 2012, and thus provide a comprehensive view of the adoption dynamics associated with this cutting-edge technology for sustainable agricultural intensification and its other backward and forward linkages value chain actors.

Improving our understanding of IAA value chain participation dynamics is timely and important for a number of reasons. Firstly, IAA value chain participation dynamics research is very relevant to overall systems technology adoption and diffusion. Most existing technology adoption research, including systems technology adoption studies, overlooks the prevalence of dis-adoption and the reasons behind it. This may be due to a lack of information on technology use over time. Secondly, studies on IAA value chain participation dynamics will add momentum to the sustainable agricultural intensification technology adoption and diffusion paradigm, which is an important issue in agricultural development policy agendas in developing countries. Finally, relatively little empirical work has been done to examine the factors that impede or facilitate the adoption and diffusion of system technologies, especially IAA. A better understanding of the constraints on farmer adoption and diffusion behaviour is therefore important for designing more effective pro-poor policies intended to stimulate the adoption of system technologies and increase productivity.

The research presented in this chapter contributes to the growing body of technology adoption literature, particularly on sustainable agriculture technology, in

---

13 A brief description of IAA in Bangladesh is given in the next section, for a detailed review see Prein (2002), Edwards (1998), and Pant et al. (2005).

a number of ways. Firstly, the analysis was based on a large and comprehensive three-wave household panel dataset derived from a survey conducted from 2007–2012 on IAA value chains among indigenous households in Bangladesh. Secondly, we integrated the value chain approach with technology adoption by considering all of the actors along IAA value chains. Thirdly, the panel data permitted the analysis: (i) to extend the static adoption framework to a dynamic one by identifying the factors that distinguish among non-participators, participators, and dis-participators[14]; (ii) to control for unobserved heterogeneity (omitted variables bias that causes biased and inaccurate estimation results) in order to answer the question "what happens to participation status over time?" and verify the reliability of other estimates; and (iii) to use lagged measures of key explanatory variables to control for potential endogeneity of participation determinants. Fourthly, the use of this method recognizes the sequential versus simultaneous nature of some farm-level decisions concerning technology adoption or other productive strategies (Barham et al., 2004). As Wu and Babcock (1998) found, without controlling for technology interdependence and simultaneous adoption in complex farming systems it is possible to either underestimate or overestimate the effects of various factors on technology adoption decisions. Barham et al. (2004) used all of these methods to study the dynamics of agricultural biotechnology adoption in the state of Wisconsin (USA), but here we used these methods to study IAA value chain participation dynamics. Fifthly, to check the robustness of the results we used a correlated random effect (CRE) model, which is an extension of the approach by Barham et al. (2004) that allows for correlation between time invariant unobserved individual heterogeneity and the observable variables.

In the next section, after explaining the dynamics of the IAA value chain participation in the broader development context in Bangladesh, we briefly review the literature on the issues of dynamics, omitted variables, and endogenous regressors in the context of previous IAA adoption studies. In the following section, our empirical econometric modelling approaches, multinomial logit and several panel methods are applied to the dynamic process of IAA value chain participation and dis-participation are discussed. In the fourth section, the panel data and some descriptive statistics are presented that support the econometric analysis of the dynamics of IAA value chain adoption/participation. In the sixth section, we discuss the empirical model results on factors explaining IAA value

---

14 Very few studies have examined abandonment (dis-adoption/dis-participation) with respect to technology adoption dynamics (Carletto et al., 1996; Neill and Lee, 2001; Moser and Barrett, 2003, 2006; Barham et al., 2004; Bravo-Ureta et al., 2006; An, 2008; Läpple, 2010).

chain participation dynamics. Finally, we summarize the chapter's empirical results on the participation dynamics of IAA both in the context of extremely poor marginalized households in Bangladesh and in terms of general approaches for the promotion of sustainable development practices along with the limitations of this effort and further research options.

3.2 Integrated Aquaculture-Agriculture Value Chain Participation Dynamics: Issues and Approaches

3.2.1 Integrated Aquaculture-Agriculture in Bangladesh

The general idea behind IAA systems is to integrate two or more farming systems to increase system productivity and efficiency through the synergistic effects among the subsystems (Prein, 2002; Dey et al., 2010; Jahan and. Pemsl, 2011). Given the many socio-economic and environmental problems in the Bangladeshi agricultural sector and recognizing the broad potential of aquaculture (e.g. improved nutrition, employment, foreign exchange earnings, positive environmental benefits, etc.,); different governmental, non-governmental, private, and international organisations have been promoting IAA systems as a sustainable intensification option for small-scale farmers (Karim, 2006; Jahan and Pemsl, 2011). IAA systems range from simple to complex systems with many sources and linkages, particularly in a smallholder context. Like many other Asian and recently African countries, two basic types of IAA systems exist in Bangladesh.[15] The first integrates typically small, seasonal ponds into an existing agricultural production system and receives nutrient inputs such as crop residues and other by-products such as poultry-fish and duck-fish based IAA systems (Ahmed, 1992; Gupta et al., 1992; Ali et al., 1995; Azim and Wahab, 1998; Samsuzzaman, 2002; Jahan and Pemsl, 2011). The second approach is where an aquaculture system is physically integrated into another farming system through modification of the design and operation of the latter, such as fish or prawn production in rice fields (Haroon et al., 1992; Ahmed and Garnett, 2011; Ahmed et. al., 2011; Jahan and. Pemsl, 2011). Both IAA systems have greater potential and practical applications in Bangladesh, but their adoption has been marginal (Haroon et al., 1992; Karim, 2006; Dey et al., 2007;

---

15 See Buck et al. (1979), Pullin and Shehadeh (1980), Sharma and Olah (1986), Little and Muir (1987), Ruddle and Zhong (1988), Guan and Chen (1989), Edwards (1993), Symones and Micha (1995), Mathias et al. (1998), Rothuis et al. (1998), Dalsgaard and Prein (1999), FAO (2001), Prein (2002), and Dey et al. (2010) for examples of different types of IAA in Bangladesh, China, India, Indonesia, Malaysia, Thailand, and Vietnam in Asia, Ghana and Malawi in Africa, and Hungary and Germany in Europe.

Nhan et al., 2007; Ahmed and Garnett, 2011; Ahmed et. al., 2011). Some estimates suggest that Bangladesh has between two and three million hectares that are suitable for integrated rice–fish production (ADB, 2005; Dey and Prein, 2006; Ahmed and Garnett, 2011), but recent survey results indicate that only about 0.18 million hectares are currently under integrated rice–fish systems (Dey et al., 2013).

A satellite survey conducted by the DOF, Fisheries Resources Survey System (FRSS) indicated that there are 1.95 million small ponds in Bangladesh (Huda et al., 2010), however, a later survey found 182% more ponds with an cumulative area that is 78% greater than the total reported by FRSS (Belton and Azad, 2012). Approximately 4.27 million households reported having a homestead pond. In addition there are 114942 ha of commercially operated ponds (Belton and Azad, 2012). Although there are different national and international organisations and NGOs promoting IAA, particularly integrated rice–fish based IAA systems, through various projects such as the Research for Sustainable Aquaculture Development Project; the Development of Sustainable Aquaculture Project, the AFP of WF, Greater options for local development through aquaculture, and the 'New options for pest management' project of CARE, however, the adoption rates of these innovative systems are often low and not uniform. Thus to achieve its full potential it is necessary to better understand the factors that impede or facilitate the adoption and diffusion of IAA in Bangladesh.

3.2.2 Integrated Aquaculture-Agriculture Adoption Research Issues

IAA studies from Bangladesh and elsewhere appear to have been based soley on cross-sectional rather than panel data, and studies on IAA adoption are rare, particularly micro-econometric studies. A very recent study on integrated rice–fish production IAA in Bangladesh stated that IAA studies primarily focus on biological and technical feasibility rather than feasibility at a system level. Socioeconomic, policy, and institutional dimensions of integrated rice–fish production systems research is lacking (Dey et al., 2013). Most adoption studies are based on descriptive statistics and all micro-econometric studies have only considered classic analysis of the determinants of adoption as opposed to the decision not to adopt. Although it is clearly understood that some farmers were dis-adopting, none of these studies considered the dynamic issues that arise from the relatively high degree of 'dis-adoption' of IAA in Bangladesh and elsewhere. But it is very important to known about the reasons for 'dis-adoption' or about farmers that continue adopted practices (i.e. who are successful with the method for assessing claims that the technology would greatly benefit poor smallholder farmers). Unfortunately and perhaps surprisingly, this aspect is commonly absent from almost all sustainable technologies adoption studies (Lee and Ruben, 2000). In addition, none of the IAA studies we found had considered the value chain approach for taking into account the backward and forward linkage IAA value chain actors.

Cross-sectional estimates of IAA adoption/participation may also suffer from omitted variables (unobserved heterogeneity) bias. In particular, a number of unobservable farm and individual characteristics (e.g. value chain actor management ability or motivation, learning from previous experience with IAA, and neighbourhood learning effects) that are correlated with observed variables used as regressors may influence the profitability of IAA activities and hence adoption/participation decisions. If these omitted variables significantly influence the benefits or costs of IAA value chain participation it would lead to inconsistent parameters estimates in the participation equation. In addition, some covariates may be endogenous to the IAA value chain participation decision. To correct for some of the problems arising from the use of cross-sectional survey data, we used panel data and several relevant econometric methods such as a Random Effects (RE) logit model and a CRE logit model to control for fixed and random missing variables effects (Wooldridge, 2010a). With cross-sectional data a typical solution for dealing with potential endogeneity problems in explanatory variables is to find suitable instrumental variables that are plausibly correlated with the participation decision, but that are uncorrelated with the error term. However, panel data provide a straightforward way to control for endogeneity associated with explanatory variables and their influence on the participation decision. We used lagged independent variables as instruments that are unlikely to be correlated to the dependent variable, but may be correlated with the endogenous explanatory variable. Similar approaches were used to control for unobserved heterogeneity (missing variables) and potential endogeneity problems in research on recombinant bovine somatotropin (rBST) adoption dynamics among Wisconsin farmers by Barham et al. (2004). Here we extended these approaches by using an additional CRE model.

3.3 Empirical Econometric Estimation Framework

3.3.1 Conceptual Framework for Integrated Aquaculture-Agriculture Value Chain Participation

As discussed above in section 3.2, IAA value chain adoption or participation decisions are dynamic and several issues arise in the estimation of model parameters. Following the approaches introduced in Barham et al. (2004) we used three IAA value chain participation models to address different estimation issues. First we used a multinomial logit model that separates individual households into three distinct participation categories by identifying the factors that distinguish them. In the second estimation we used the panel structure of the data to control for unobserved variables in a RE and a CRE logit models as a means of testing the consistency of the multinomial logit results of the first estimation. Finally, we re-specified the RE logit panel model using lagged regressors as instruments to control for endogeneity.

Our objective was to identify the determinants of IAA value chain participation. The decision of whether or not to participate in IAA value chain is associated with potential costs and benefits, which may be perceived differently by individual households. Following the standard index model (see Wooldridge, 2010b), individual household decisions on IAA value chain participation can be modelled using a random utility framework (Rahm and Huffman, 1984; Feder et al., 1985; Marra et al., 2003) or can be modelled by the relative costs and benefits of the IAA value chain participation relative to alternative strategies. These two alternative formulations have slightly different motivations, however, that lead to the same set of econometric estimation techniques (Barham et al., 2004).

According to this framework, the actual utility or benefit level of IAA value chain participation to each household in each time period is unknown. However, the household chooses to participate in the IAA value chain if the perceived utility or benefit gained is greater than the utility or benefit of non-participation. Thus the marginal benefit or utility of IAA value chain participation can then be expressed as a function of observed characteristics (Z) in the latent variable model as follows:

$$C^*_{it} = \beta Z_{it} + \varepsilon_{it} \tag{1}$$

where C^*_{it} is an indicator of the latent IAA value chain participation status, β is a vector of parameters to be estimated that can assume different values according to the time period, and ε_{it} is the disturbance term that may have both specific and idiosyncratic individual household elements. In turn, the observed dependent variable, which depicts IAA value chain participation status during each time period (C_{it}), where $C_{it} = 1$ for participation and $C_{it} = 0$ for non-participation relative to C^*_{it} as follows:

$$C_{it} = \begin{matrix} 1 \text{ if } C^*_{it} > 1 \\ 0 \text{ if } C^*_{it} \leq 0 \end{matrix} \tag{2}$$

The probability of IAA value chain participation was described as a function of the observed explanatory variables as:

$$\Pr(C^*_{it} > 0) = \Pr(C_{it} = 1) = F(\beta'Z_{it}) \tag{3}$$

where $F(\beta'Z_{it})$ is a function that can be either logistic or normal. The logistic function was chosen in order to make consistent assumptions across the different model results discussed below following Barham et al. (2004). The choice of explanatory variables included in Z is directed by empirical literature on integrated rice–fish based IAA technology adoption (e.g. Prein, 2002; Nabi, 2008; Ahmed and Garnett, 2010; Jahan and Pemsl, 2011; Bosma et al., 2012).

3.3.1.1 Multinomial logit model

As mentioned earlier, over time there can be at least three categories of participation behaviour: 'participators,' 'non-participators,' and 'dis-participators.' There could be fundamental differences among these three categories that reveal more about the dynamics of the participation decisions than a simple cross-sectional binary analysis allows. Panel multinomial logit model can capture the distinct determinants of these participator categories. Specifically, the determinants associated with each participator category can be juxtaposed with those in a base category (McFadden, 1974; Zepeda, 1994; Barham, 1996, Barham et al., 2004).

Using the formulation above the different unordered outcomes can be described by the notations $j = 0, 1...J$, and time $t = 0, 1... T$, then the categories j can be defined as:

$j = 0$ if $C^*_{it} \leq 0$ $\forall t$, Non-participation
$j = 1$ if $C^*_{it} > 0$ $\forall t$, Participation
$j = 2$ if $C^*_{it} > 0$ for $t = s$, but $C^*_{it} \leq 0$ for $t = T$, Dis-participation

Although the formulation above is dynamic, if the characteristics that determine which category a farmer belongs to can be sufficiently defined in the first time period ($Z_{it} = 0$) the problem can be reduced to a single period estimation. Dropping time subscripts and using the notation above, the standard approach to a multinomial logit relies on the Weibull distribution for the various disturbance terms. Then, the J+1 unordered outcomes occur with a probability determined by the following equation.

$$\text{Prob}(C_i = j) = \frac{e^{\beta'j Z_i}}{\sum_0^2 e^{\beta'k Z_i}}, \quad j = 0, 1, 2. \tag{4}$$

In order to identify the J+1 possible unordered outcomes in the model and the model parameters, a standard normalization was used to assign a benchmark outcome so that the parameter matrix $\beta_0 = 0$. This technique allows the rest of the coefficients in the estimation of the different choices to be identified relative to the base outcome. The multinomial logit model, as specified, also partially addresses the endogeneity issue by using explanatory variables from the baseline to describe the IAA value chain participation process from the baseline until the current period (2007–2012). While it does not remove potential endogeneity from the estimates for participators, the baseline farm and household characteristics are plausibly exogenous for all other categories of farmers who made the decision to participate or dis-participation afterwards. This solution to the endogeneity issue comes at a cost, which is that the model describes these decisions made in 2012 as being a function of farm and household characteristics fixed at 2007 values (Greene, 2003; Barham et al., 2004). The random effects logit model discussed below relaxes this stricture.

3.3.1.2 Random Effects logit model

Following Barham et al. (2004), Greene (2003), and Guilkey and Murphy (1993) we used a RE logit panel data model to account for omitted variables and the possible endogeneity of some regressors. The latent model of IAA value chain participation can be specified as:

$$C^*_{it} = \beta Z_{it} + \mu_i + \varepsilon_{it} \quad i = 1, 2 \ldots N; t = 1 \ldots T \tag{5}$$

where C^*_{it} is a latent dependent variable; C_{it} is the observed binary outcome variable defined as

$$C_{it} = \begin{cases} 1 & \text{if } C^*_{it} > 0; \\ 0, & \text{otherwise.} \end{cases} \tag{6}$$

Z_{it} represents a vector of time-varying and time-invariant exogenous variables that influence C^*; β represents a vector of parameters to be estimated; μ_i is a term that captures unobserved individual (household in this case) heterogeneity, and ε_{it} is a random error term. The subscripts i and t refer to households and time periods respectively. The term μ_i is distributed normally with a mean of zero and a variance, σ^2_μ, and ε_{it} has a logistic distribution with a mean of zero and a variance, σ^2_ε (Barham et al., 2004). The RE logit model was estimated using the xtlogit command in Stata 12.1.

3.3.1.3 Re-specified Random Effects logit model for addressing endogeneity

If a subset X_{it} of the independent variables Z_{it} are endogenous to the dependent variable C^*_{it} then the independent variables Z_{it} will be correlated with the error term ε_{it}. This condition violates one of the model's basic assumptions. As Barham et al. (2004) stated in the context of a panel data model, lagged independent variables make reasonable instruments in that they are likely to be uncorrelated to the dependent variable, but correlated with the contemporaneous dependent variable of interest. Thus if some portions of independent variable Z_{it} are a set of exogenous variables, $\hat{Z}_{it}$, and potentially endogenous variables X_{it}, the following model can be expressed using lagged endogenous variables:

$$C^*_{it} = \beta \hat{Z}_{it} + X_{it-1} + \mu_i + \varepsilon_{it}$$

Following Barham et al. (2004) this model was estimated using the same procedure used in the RE logit model described above with the replacement of potentially endogenous variables with their lagged counter parts.

3.3.1.4 Correlated Random Effects model

RE models assume that the time invariant unobserved individual heterogeneity, μ_i, is uncorrelated with the observable variables Z_{it}, $\forall i$, and t (strict exogeneity assumption). However, it is unrealistic in many cases that the unobserved heterogeneity will be orthogonal and uncorrelated to the other covariates. For example, the unobserved variables motivation or ability, which are captured by αi, would be correlated with some of the observed regressors such as education, which in turn introduce bias in the coefficient estimates. This problem can be solved by using a Fixed Effects (FE) model, which cancels out the individual effects in the estimation process. A FE model cannot be used here because of the incidental parameters problem in nonlinear models. However, following Mundlak (1978) and Chamberlain (1984) CRE nonlinear panel data models relax the strict exogeneity assumption by allowing for correlation (between μ_i and X_{it}). CRE models are estimated by adding the mean of the time-varying x-variables as an additional set of regressors in the model. The inclusion of these additional mean variables controls for time constant unobserved heterogeneity (Wooldridge, 2010a; Alem et al., 2013; Bezu et al., 2014). The RE equation is estimated using the CRE estimator.[16] The CRE model was estimated by using the mundlak full command in Stata 12.1 (Perales, 2013).

3.4 Data and Descriptive Statistics

The analyses presented in this chapter are based on three-wave (2007, 2009 and 2012) panel survey data on IAA value chain participation among indigenous households in Bangladesh (see Table 1.1). The panel data are unbalanced, but here only balanced panel data sets were used because the dynamics of the IAA value chain participation process are of interest. Only the baseline (2007) survey responses were used for the multinomial logit analysis. On the other hand, panel data

---

16 The CRE approach was also used to control for heterogeneity by Chernozhukov et al. (2005), Bester and Hansen (2007), Papke and Wooldridge (2008), and Weidner (2011).

from three observations periods (2007, 2009, and 2012) for 570 households were used for the RE logit analysis. The panel method estimates control for unobserved heterogeneity and any potential endogeneity problems. Specifically, the re-specified RE regression model controls for endogeneity by using the lagged variables as instruments and the CRE regression model controls for unobserved heterogeneity.

Table 3.1 shows the characteristics of the first round, baseline data used in this study. The continuous 'participators' category accounted for 41.05% of the sample (25.44% of the total sample were involved in IAA production and 15.61% participated in up and downstream IAA value chain activities). This includes IAA value chain participants from the beginning of the panel study in 2007 until 2012. The 'dis-participators,' category accounted for 37.89% of the sample (9.65% of the total sample were involved in production and 28.25% participated in up and downstream value chain activities). The respondents in this category had been value chain participants in 2007 but had stopped by 2012. The 'non-participators' group, which accounted for 21.05 % of the sample, did not participate in IAA value chain activities. There were nearly as many 'dis-participators' (38%) as there were continuous 'participators.' Of the entire sample, the proportion of 'dis-participators' that had been involved in up and down stream IAA value chain activities was nearly three times greater than the proportion that were 'dis-participators.'

The descriptive statistics presented in Table 3.1 provide a profile of the sample characteristics of the different IAA value chain participation categories in 2007 based on IAA value chain participation status in 2012. Table A.3.1 in the Appendix presents the results of the t-test comparison of the means of selected variables among participation categories for all value chain actors. Some of these characteristics are the explanatory variables used in the models featured in subsequent analyses. Household head education was greater for IAA production 'participators' and 'dis-participators' than for 'non-participators.' IAA value chain 'participators' and 'dis-participators' were also more likely to have access to market information related to price and farm technology. The mean age of the household head was higher for 'dis-participators' and 'non-participators' than for 'participators.' IAA value chain production 'participators' and 'dis-participators' were also distinguishable in terms of household assets, poultry and livestock ownership, CBO membership status, and access to extension services. Moreover, among both IAA value chain production 'participators' and 'dis-participators' there were a greater number of household heads that were married, male, and that worked mainly in the agricultural sector. The participator groups were also significantly different in terms of welfare, measured in terms of farm income from non-fisheries produce. Similarly, farm and family sizes of IAA production 'participators' were higher than for 'non-participators.'

Table 3.1: Mean and standard deviations of independent variables by integrated aquaculture-agriculture value chain participation category in Bangladesh

Variable	Definition and measurement	Total sample	Participation in up and down stream value chain activities		Participation in value chain production activities		Non-participators
			Participators	Dis-participators	Participators	Dis-participators	
Education of HH head	Schooling of household head in years	3.22 (3.85)	3.13 (3.75)	2.23 (3.15)	4.22 (4.14)	3.44 (4.22)	3.33 (3.98)
Occupation of HH head	=1 if the main occupation of the HH head is agriculture	0.39 (0.49)	0.16 (0.37)	0.20 (0.40)	0.61 (0.49)	0.65 (0.48)	0.44 (0.50)
Marital status of HH head	=1 if the HH head is married	0.92 (0.28)	0.92 (0.27)	0.92 (0.27)	0.96 (0.20)	0.89 (0.31)	0.87 (0.34)
Gender of HH head	=1 if HH head is male	0.94 (0.23)	0.96 (0.21)	0.94 (0.24)	0.97 (0.18)	0.95 (0.23)	0.91 (0.29)
Age of HH head	Age of HH head in years	44.72 (12.13)	37.34 (8.67)	42.66 (11.32)	47.37 (12.20)	49.71 (14.18)	47.47 (11.40)
HH size	Total number of HH members	4.52 (1.59)	4.20 (1.30)	4.17 (1.41)	5.03 (1.71)	4.75 (1.85)	4.51 (1.57)
Residence size	Total number of rooms	1.51 (0.66)	1.30 (0.51)	1.40 (0.57)	1.74 (0.74)	1.85 (0.76)	1.38 (0.59)
Assets†	Total number of assets	6.34 (5.52)	4.45 (3.27)	3.91 (3.08)	9.44 (6.20)	7.84 (6.01)	6.57 (6.23)
Farm size	Total land area in decimals	106.21 (123.4)	37.20 (41.33)	54.00 (67.63)	173.52 (146.22)	171.74 (161.21)	116.09 (113.32)
Poultry and livestock	Total number of poultry and livestock	11.52 (11.75)	11.25 (10.41)	10.55 (10.92)	14.38 (16.00)	11.05 (9.39)	9.76 (7.47)
Fisheries Income	Annual fisheries and related income (in BDT)	1400.48 (3879.79)	3089.01 (7993.56)	1325.37 (2094.94)	980.52 (3146.90)	825.55 (2050.18)	1019.90 (1509.27)
Farm Income	Annual agricultural and related income (in BDT)	23106.26 (26876.0)	10169.36 (18433.80)	11858.05 (16352.54)	32442.82 (27372.91)	34668.62 (33492.83)	31211.41 (29943.86)
Non-farm Income	Annual non-agricultural income (in BDT)	21530.41 (16282.3)	21551.73 (11937.01)	24748.57 (12594.02)	18411.83 (17907.34)	21025.09 (21227.33)	21196.79 (18177.29)
CBO membership	=1 if the household head is a member of a CBO	0.85 (0.35)	0.99 (0.11)	0.98 (0.14)	0.98 (0.14)	0.98 (0.13)	0.38 (0.49)
Access to extension services	=1 if household has access to extension services	0.94 (0.25)	0.90 (0.30)	0.91 (0.29)	0.98 (0.14)	0.95 (0.23)	0.94 (0.24)
Access to irrigation	=1 if irrigated crops in previous year	0.63 (0.48)	0.47 (0.50)	0.49 (0.50)	0.77 (0.43)	0.80 (0.40)	0.71 (0.46)
Access to credit	= 1 if able to access credit	0.92 (0.28)	0.92 (0.27)	0.90 (0.30)	0.92 (0.27)	0.89 (0.31)	0.93 (0.25)
Access to market information	=1 if agricultural market information available	0.84 (0.37)	0.72 (0.45)	0.75 (0.44)	0.97 (0.18)	0.93 (0.26)	0.84 (0.37)
Number (percentage) of samples		570 (100)	89 (15.61)	161 (28.25)	145 (25.44)	55 (9.65)	120 (21.05)

Notes: Standard deviations are shown in parentheses; †assets include agricultural assets (e.g., shallow machines, insecticide sprayers); productive assets (e.g. tractors, rickshaws, vans, cars, ploughs, threshers, sewing machines, carts); durable consumer goods (e.g. televisions, radios, motorcycles, bicycles, mobile phones, clocks); and residential assets (e.g. beds, chairs, tables, hand tubewells).

The summary in Table 3.1 also reveals that the differences between the two 'participator' categories (including 'dis-participators') are less than the differences between 'participators' and 'non-participators.' Farm and non-farm income are comparatively higher for 'dis-participators' than for 'participators.' Similarly, mean household head age and irrigation access were greater among 'dis-participators' than for 'participators' and 'non-participators.' The mean values of other variables such as household head education, family size, farm size, and income from fisheries were distinct between 'dis-participators' and 'participators'.

Unlike the IAA production segment actors, IAA up and downstream 'participators' and 'dis-participators' were distinctly opposite from 'non-participators' with respect to many variables. Up and downstream IAA value chain 'participators' had household heads that were less educated, younger, less engaged in agriculture, had smaller families and farms, earned less farm income, had fewer assets, and had less access to irrigation, extension services, and market information than 'non-participators,' but had greater fisheries and non-farm income as well as more poultry and livestock than 'non-participators.'[17] Up and down stream IAA value chain activity 'participators' and 'dis-participators' also differed in terms of household head education and age; farm size; farm, non-farm and fisheries income; and access to irrigation and market information. Overall, the sample socio-economic characteristics reveal that up and downstream IAA value chain actors were comparatively worse off than IAA production actors.

Table 3.2 presents the reported reasons for dis-participation from IAA value chain by various actors. The three top ranking cited problems were the scarcity of fish, water, ponds, lack of capital, and insufficient income from IAA value chain participation. All other reported reasons for discontinuing IAA participation were relatively insignificant. The top two reasons, income and resource scarcity, are associated with one another, which is also apparent from the relatively low fisheries income of 'dis-participators' relative to 'participators' presented in Table 3.1. Although 'dis-participators' incomes were lower than those of 'participators' they were higher than 'non-participators.'

---

17 Sample household farm size and annual household income were also very low compared to the mean national farm size (0.62 ha or 153.14 decimal) and household annual income (137760 BDT) (BBS, 2011; Krishi diary, 2011).

Table 3.2: Reported reasons for dis-participation from integrated aquaculture-agriculture value chains in Bangladesh

Reason for dis-participation/adoption	Frequency†	Percentage
Lack of capital	52	16
Lack of labour	24	8
Not or less profitable/income	35	11
Scarcity of fish, water, or pond	61	19
Scarcity of necessary resources (like cage materials)	32	10
Risk of losses from predators and theft	16	5
Flood and drought	5	2
Lack of suitable land	12	4
Lack of technical knowledge	2	1
No answer	81	25
Sample size	216	

Note: † Multiple responses were provided by some households are counted, therefore the percentages do not sum to 100.

3.5 Econometric Results

3.5.1 Multinomial Logit Model Results

Table 3.3 presents the results of the multinomial logit regression. IAA value chain 'participators' and 'dis-participators' were different from 'non-participators' (base category) in both the production (columns 2 and 3) and non-production value chain segments (columns 5 and 6). The table also shows that 'dis-participators' (columns 4 and 7) differ from 'participators' (base category) for both value chain participation categories.

Asset ownership, CBO membership, farm income and market information access are significant predictors of IAA production process participation. Households that participated in the IAA production reported more assets than non-participants. The CBO membership and access to market information coefficients have similar signs and were significant, which indicates IAA production process 'participators' have significantly higher levels of CBO membership and access to market information than 'non-participators.' Household farm income was negatively associated with participation in IAA production activities. Similarly, farm sizes were larger and CBO membership greater among 'dis-participators' relative to 'non-participators.' In Table 3.3 column 4 lists the values for 'dis-participators' compare to 'participators' and shows that increase in household assets was negatively associated with IAA production process dis-participation relative to participation.

Table 3.3: *Multinomial logit analysis results of integrated aquaculture-agriculture value chain participation in Bangladesh*

Variables	Production process participation			Upstream and downstream participation		
	Participants	Dis-participators	Dis-participators†	Participants	Dis-participators	Dis-participators†
1	2	3	4	5	6	7
HH head education	−0.02 (0.05)	−0.03 (0.06)	−0.01 (0.05)	−0.02 (0.06)	−0.08 (0.06)	−0.06 (0.05)
HH head gender	0.86 (0.88)	0.40 (0.96)	−0.46 (0.83)	0.51 (0.95)	0.07 (0.83)	−0.44 (0.65)
HH head age	−0.01 (0.02)	0.01 (0.02)	0.02 (0.01)	−0.09*** (0.02)	−0.05*** (0.02)	0.05*** (0.02)
HH size	−0.03 (0.11)	−0.11 (0.13)	−0.08 (0.11)	−0.20 (0.17)	−0.32** (0.15)	−0.12 (0.12)
Number of Assets	0.10** (0.04)	0.03 (0.05)	−0.07* (0.04)	0.03 (0.06)	−0.06 (0.05)	−0.09* (0.05)
Farm size	0.00 (0.00)	0.00** (0.00)	0.00 (0.00)	−0.01** (0.00)	−0.00 (0.00)	0.01** (0.00)
Poultry and livestock	−0.00 (0.02)	−0.02 (0.02)	−0.02 (0.02)	0.01 (0.02)	0.01 (0.02)	−0.01 (0.01)
Fisheries income	−0.00 (0.00)	−0.00 (0.00)	−0.00 (0.00)	0.18* (0.10)	0.09 (0.09)	−0.09** (0.05)
Farm income	−0.00*** (0.00)	−0.00 (0.00)	0.00 (0.00)	−0.030* (0.01)	−0.03*** (0.01)	0.00 (0.01)
Non-farm income	−0.00 (0.00)	0.00 (0.00)	0.00 (0.00)	−0.01 (0.02)	0.01 (0.01)	0.03** (0.01)
CBO membership	4.48*** (0.66)	4.68*** (1.05)	0.21 (1.19)	5.21*** (1.08)	4.84*** (0.67)	−0.37 (1.18)
Access to extension services	0.58 (0.88)	0.18 (0.90)	−0.39 (0.89)	0.26 (0.76)	0.31 (0.68)	0.06 (0.53)
Access to irrigation	0.08 (0.47)	0.56 (0.58)	0.48 (0.48)	0.07 (0.56)	−0.00 (0.47)	−0.07 (0.40)
Access to credit	−0.51 (0.78)	−1.04 (0.83)	−0.53 (0.60)	−0.24 (0.88)	−0.37 (0.75)	−0.14 (0.56)
Access to market information	1.73** (0.72)	0.64 (0.78)	−1.09 (0.80)	0.44 (0.61)	0.35 (0.55)	−0.09 (0.39)
Constant	−5.17*** (1.87)	−4.91** (2.14)	0.26 (2.05)	0.33 (1.88)	0.35 (1.54)	0.02 (1.52)
Number of observations		320			370	
LR chi²(30)		196.83			287.2	
Log likelihood		−230.92			−252.304	
Pseudo R²		0.2988			0.3627	

Notes: Categories in 2012 are based on 2007 characteristics; the 'non-participator' category is the comparison group except for the column indicated with†, for which the 'participator' group was used as the comparison group; standard deviations are indicated in parentheses; * significance at 10%, ** significance at 5%; *** significance at 1%.

Similarly, columns 5 and 6 in Table 3.3 show the determinants of up and down stream value chain 'participators' and 'dis-participators' compared to 'non-participators.' The last column (7) shows the determinants of 'dis-participators' compared to 'participators.' The age of the household head was negatively and significantly associated with up and downstream IAA value chain 'participators' and 'dis-participators.' This may be because as the age of the household head increases the likelihood of participation decreases. Similarly, family size was negatively associated with dis-participation, which suggests that an increase in family size would decrease the probability of dis-participation relative to non-participation. Asset ownership was also negatively associated with dis-participation compared to participation (column 7). Farm size was negatively associated with IAA up and down stream participation. Farm size was positively associated with dis-participation relative to participation. Fisheries income was positively associated with up and down stream IAA value chain participation relative to non-participation, and negatively to dis-participation relative to participation. On the contrary, increased farm income was negatively associated with participation and dis-participation relative to non-participation. As expected, increased non-farm income was positively associated with dis-participation relative to participation among up and downstream IAA value chain actors. The relationship of CBO membership with participation and dis-participation in up and down stream IAA value chain activities was the same as for IAA production related activity actors.

3.5.2 Random Effects Logit Model Results

The second and fifth columns in Table 3.4 show the results of a RE logit estimation of the probability that a household participates in IAA value chains in each year (2007, 2009, 2012) as a function of the farm and farmer's characteristics in that year. The estimates of the unobserved random variable (ρ) are significantly different from zero, justifying the use of a RE model and attention to unobserved variables in the IAA value chain participation estimation. Unlike the results of the multinomial logit model, education had a positive and statistically significant relationship with participation in the both production and non-production segments of the IAA value chain. Continued participation in the up and down stream IAA value chain activities had a significant negative relationship with household head age, which is similar to the multinomial model results, but for production segment participation there were not any significant relationships in the RE model results. In contrast to the multinomial results, household size was positively associated with continued participation in IAA value chain activities, and this relationship was highly significant for production segment participators. Farm size had a negative association on up and down stream IAA value chain participation, which is consistent with the multinomial model results. Conversely, income from fisheries had a significant positive association with continued participation in IAA value chains. This result is consistent with IAA up and down stream participation results of the multinomial model. Contrary to the multinomial results, however, farm income had a significant positive relationship with up and downstream IAA value chain participation. CBO membership had a positive relationship with IAA value chain participation similar to the multinomial model results. Likewise, access to extension services and market information were significantly and positively associated with participation, which is also similar to the multinomial model results except that only access to market information was significantly positive for participation in the IAA production activities in those results. The RE model results suggest that access to extension services and market information have a greater influence after controlling for unobserved household and farm-specific factors.

In Table 3.4 the third and sixth columns show the CRE model results, which allow correlation between some of the explanatory variables and the unobserved individual heterogeneity term. The sign and significance of the coefficients are similar to the RE logit regression results, while the magnitudes of the coefficients are slightly lower than the RE model results. The fourth and seventh columns present the lagged RE logit model results with four potentially endogenous variables: asset ownership, poultry and livestock ownership, fisheries income, and farm income, lagged by one period. Only two of the panel datasets were used, 2009 and 2012, along with the lagged variables for 2007 and 2009. The results are quite similar to the RE logit regression that only used contemporaneous explanatory variables. Overall the consistency of these core results suggests that the relationships of household head education, family size, CBO membership, and access to extension services and market information to IAA value chain participation are robust. To test for the robustness of the different participation determinants we compared the results of the three alternative nonlinear specifications: the RE logit, CRE logit, and the Lagged RE logit models. The results (in Table 3.4) for all of the model specifications are similar, indicating robustness.

Table 3.4: Random Effects model results of IAA value chain participation

Variables	Production process participation			Upstream and downstream participation		
	Random Effects Logit	Correlated Random Effects (CRE) Model†	Lagged Random Effects Logit	Random Effects Logit	Correlated Random Effects (CRE) Model†	Lagged Random Effects Logit
1	2	3	4	5	6	7
HH head education	0.08***	0.02***	0.08***	0.08***	0.02***	0.09***
	(0.02)	(0.00)	(0.02)	(0.02)	(0.00)	(0.02)
HH head gender	0.04	0.01	0.12	0.17	0.03	0.31
	(0.26)	(0.05)	(0.29)	(0.26)	(0.05)	(0.29)
HH head age	-0.00	-0.00	0.00	-0.02***	-0.00***	-0.01
	(0.01)	(0.00)	(0.01)	(0.01)	(0.00)	(0.01)
Household size	0.11***	0.02**	0.17***	0.04	0.00	0.08***
	(0.04)	(0.01)	(0.05)	(0.05)	(0.01)	(0.05)
Asset ownership	-0.00	-0.00		-0.00	-0.00	
	(0.00)	(0.00)		(0.00)	(0.00)	
Asset ownership in previous period			0.04***			0.01
			(0.01)			(0.02)
Farm size	-0.00	-0.00	-0.00	-0.00**	-0.00**	-0.00**
	(0.00)	(0.00)	(0.00)	(0.00)	(0.00)	(0.00)
Poultry and livestock ownership	0.00	0.00**		0.00	0.00***	
	(0.00)	(0.00)		(0.00)	(0.00)	
Poultry and livestock ownership in previous period			0.00			0.00
			(0.00)			(0.00)
Fisheries income	0.01***	0.00***		0.01***	0.00***	
	(0.00)	(0.00)		(0.00)	(0.00)	
Fisheries income in previous period			0.00			0.00
			(0.00)			(0.00)
Farm Income	0.00	0.00**		0.00**	0.00***	
	(0.00)	(0.00)		(0.00)	(0.00)	
Farm income in previous period			-0.00			0.00
			(0.00)			(0.00)
Non-farm income	0.00	0.00	-0.00	0.00	0.00	-0.00
	(0.00)	(0.00)	(0.00)	(0.00)	(0.00)	(0.00)
CBO membership	1.57***	0.26***	1.41***	1.19***	0.09***	1.04***
	(0.19)	(0.03)	(0.21)	(0.20)	(0.03)	(0.21)
Access to extension services	0.56***	0.06**	0.65***	0.42***	0.02	0.54***
	(0.18)	(0.03)	(0.19)	(0.16)	(0.02)	(0.18)
Access to irrigation	0.11	0.01	0.16	-0.02	-0.02	0.10
	(0.15)	(0.03)	(0.16)	(0.15)	(0.03)	(0.16)
Access to credit	-0.20	-0.05	-0.27	-0.13	-0.03	-0.09
	(0.18)	(0.04)	(0.19)	(0.17)	(0.03)	(0.19)
Access to market information	0.64***	0.12***	0.71***	0.32**	0.05**	0.51***
	(0.16)	(0.03)	(0.17)	(0.13)	(0.03)	(0.16)
Constant	-3.00***	-0.03	-3.49***	-2.14***	0.17**	-2.76***
	(0.51)	(0.08)	(0.62)	(0.58)	(0.08)	(0.68)
Rho (ρ)	0.05***	-	0.09***	0.12***	-	0.16***
	(0.05)		(0.08)	(0.09)		(0.12)
Number of observations	1210	1210	903	1310	1310	973
Wald chi^2 (15)	146.87	194.95	111.65	113.33	151.76	74.17
Log likelihood =	-731.53	-	-545.70	-728.44	-	-566.00
Prob > chi^2	0	0	0	0	0	0

Notes: Standard deviations are shown in parentheses; * significance at 10%; ** significance at 5%; *** significance at 1%; † the time-variant variables are included as additional regressors in the CRE model, but they are not reported here to save space.

3.6 Conclusions and Policy Implications

Better understanding of the IAA value chain participation dynamics is critical for improving farming system innovations and for designing effective policy recommendations for food security and poverty alleviation in developing countries. Improved IAA technologies were introduced in Bangladesh and other parts of Asia[18] and Africa[19] by WF and its partners from national agricultural research and extension systems to provide efficient, locally suitable, low cost technologies to improve the productivity and sustainability of production among resource poor farmers who lack access to sustainable aquaculture and crop production technologies to improve incomes, household food security, and dietary nutrition. These IAA technologies include IRFFSs, fingerling production cage culture, and community-based pond aquaculture production. Supporting value chain actors considered in the analyses include fishermen, fish traders, and fingerling traders. Despite the significance of using IAA technologies in developing countries very few empirical studies have examined the factors that influence participation/adoption dynamics of these technologies value chain actors.

We examined IAA value chain production segment participation dynamics as well as the determinants of participation among other up and downstream IAA value chain actors using a range of cross-sectional and panel data econometric approaches. We focused on factors that influence IAA value chain participation and particularly dis-participation, which has received little attention in technology adoption literature. We analysed a rich and unique three-wave panel dataset collected in 2007, 2009, and 2012 from 657 indigenous farm households in northern and northwestern Bangladesh. Initially, we used the panel data to move beyond a typical comparison of participators/adopters and non-participators/non-adopters, to conduct a more nuanced multinomial logit analysis of the factors affecting 'non-participators,' 'participators,' and 'dis-participators.' The multinomial logit analysis results indicate differences between participation categories (both continuous 'participators' and 'dis-participators') and 'non-participators,' as well as differences between continuous 'participators' and 'dis-participators.' We found significant differences between 'participators' and 'non-participators,' but few differences between 'participators' and 'dis-participators,' especially among the production process 'participators' and 'dis-participators,' where the rate of

---

18 For details about the adoption and impacts of IAA in Asia (including Bangladesh) see Prein (2002), Jahan et al. (2008), and Jahan et al. (2013).
19 For details about the adoption and impacts of IAA activities in Malawi see Brummett (1999) and Dey et al. (2010).

dis-participation was lower relative to other up and downstream IAA value chain actors. We then used the panel data to estimate three RE logit models to control for fixed and random omitted variables and endogenous regressors. The results indicate significant effects of omitted variables and confirm the robustness of the results, which support the general consensus in the literature that education, family size, access to extension services, CBO membership, and access to market information are major determinants of participation in IAA value chains. The results were also found to be robust with respect to the potential endogeneity of four household characteristics.

Overall, the results of the empirical analyses of IAA participation dynamics lead to the following conclusions about future adoption and diffusion of IAA technologies in Bangladesh and other places with similar socio-economic, environmental and institutional conditions. First, the 'non-participators' appeared to be very distinct from the other participation categories. Non-participant households had a much smaller mean family size; household heads were less educated, older, and had much less access to extension services, and market information; and were less likely to be involved in a CBO. Looking ahead, 'non-participators' seem very unlikely to participate in IAA value chains without the factors or conditions that are common to value chain 'participators.' Second, the multinomial regression analysis did not reveal many distinguishing characteristics between continuous 'participators,' and 'dis-participators,' especially among actors in the production process of IAA value chains. The quantity of household assets was the only significant distinguishing characteristic between production process 'participators' and 'dis-participators.' 'Participators' and 'dis-participators' of up and downstream IAA value chain were significantly different with respect to the mean age of household heads, the quantity of assets, farm size, fisheries income, and non-farm income. Thus, younger household heads with fewer assets, larger farms, less fisheries income, and greater non-farm income were more likely to be 'dis-participators' than continue to participate in up and downstream IAA value chain activities. Yet, the current levels of CBO membership and access to extension services and market information are actually quite high among the 'dis-participators' (e.g. over 75% of up and downstream IAA value chain 'dis-participators' had access to market information), which suggests that while they may be critical determinants of IAA value chain participation, they by no means guarantee that IAA value chain participation will be profitable. Thus, changes in the levels of these factors would not be likely to make much of a difference in terms of transitioning 'dis-participators' back into value chain 'participators'.

Resource-poor farmers are capable of participating in IAA value chains, especially in non-production value chain activities. Controlling for fixed and random omitted variables and potential endogenous regressors, we found that farm size was not a key determinant of participation in either production or non-production IAA value chain activities despite often being featured as a significant determinant in technology adoption literature. This is especially possible for up and downstream value chain activity participants because they do not require land to engage in such activities. Participants in value chain production activities can operate on small parcels or rented land. This finding is robust to alternative specifications and estimation techniques. The probability of participating in IAA value chain activities increased significantly with the level of education of the household head, and together with CBO membership and access to extension services and market information, they confirm the positive relationships between technical knowledge, human capital development, and participation.

This research found that household head education; family size (used as a proxy for family labour availability); access to extension services and market information; and CBO membership are important determinants of participation in IAA value chains. Therefore policy efforts to strengthen farmer technical skills should provide platforms for active interaction among stakeholders in accordance with innovation systems theory. A good example is the effective use of broad extension and research networks, the FFS approach, and the use of existing CBOs, which facilitate interactions among individual actors. There are numerous anecdotal claims that FFS is an effective means of facilitating the transfer of technology among farmers (Röling, 2009). The findings of this study suggest that the key policy objectives for promoting IAA technology should focus on educational, institutional, and technological innovations that facilitate interactions, learning, building technical skills, and reducing labour requirements.

Although this research mostly featured household characteristics, plot/farm level or environmental characteristics and farmer attitudes regarding risk and their perceptions about technology are also important determinants of IAA value chain participation, especially for production related activities. There are increasing attempts to promote sustainable agricultural innovation and intensification options, and this study has illustrated some essential pathways. The lessons learned also have significant implications for the dissemination and durability of sustainable innovation and intensification options. Additional studies on the impact of these innovations on farmer welfare are needed to further strengthen arguments in support of sustainable innovations/intensification.

Chapter 4: Welfare Impacts of Integrated Aquaculture-Agriculture Value Chain Participation Dynamics in Bangladesh

4.1 Introduction

Despite the decreasing trends in the incidence of extreme poverty, hunger, and malnutrition in Asia this region remains home to the largest number of poor, hungry and malnourished people in the world (FAO/IFAD/WFP, 2013; ADB, 2014a). Most of these people live in rural areas furthest from roads, markets, schools, and public health services, are less likely to be educated, often belong to minority and other marginalized socio-ethnic groups, and most of them are either directly or indirectly engaged in agriculture as their primary source of livelihood (IFAD, 2003, 2011; Ahmed et. al., 2007). Markedly, agricultural intensification over the past several decades through the innovations of the GR such as high-yield seed varieties, chemical fertilisers, and modern irrigation technologies led to dramatic increases in agricultural production, livelihood improvements, and radically transformed the course of agricultural development in South and East Asia (Pender, 2007). But the impacts of the GR have also been criticized for the unintended negative long term environmental and social equity impacts and recently rice yields have been declining or remained stagnant in many parts of Asia (Pimentel and Pimentel, 1990; Pingali and Rosegrant, 1994; Kerr and Kolavalli, 1999; Das 2002; Pingali, 2012).

Like many other Asian countries, the economy of Bangladesh also largely depends on agriculture. Agriculture accounts for close to half of employment, 20% of GDP, and is the basis of food security for the entire population. Even with steady and commendable progress poverty is still widespread and continues to be largely a rural phenomenon, accounting for 84% of the nation's poor. Bangladesh also faces many challenges to food security, including but not limited to, climate change, population growth, vulnerability to price shocks, increasing natural resource scarcity, persistent poverty, and malnutrition. Most of the rural poor struggling to achieve food security are either directly or indirectly engaged with agriculture for their livelihoods, thus fostering agricultural development and sustainable rural natural resource management are crucial for reducing poverty and improving food security in Bangladesh (ADB, 2014b; Cortijo, 2014).

Like many other Asian countries rice and fish have been an essential part of Bangladeshi culture from time immemorial as household staple foods. Rice is the main source of dietary carbohydrates and fish (as well as crustaceans) as the main source of dietary animal protein (Dey et al., 2013). Rice is the leading agricultural crop, in fact so much so that in Bangladesh food security is mainly defined in terms of access to rice (Ahmed et al., 2013). The demand for rice and fish in the country is constantly increasing due to mounting population growth (Chowdhury, 2009). Like many other Asian countries rice production in Bangladesh is threatened due to land degradation (caused in part by overuse of fertilisers and pesticides), decreased arable land area, the effects of climate change, and other environmental problems (Alauddin and Tisdell, 1991; Ali et al., 1997; Rahman, 2003a, 2003b; Sarker et al., 2012). However, IAA[20] technologies potentially offer a sustainable solution to this problem by contributing to food security, income, and dietary nutrition (FAO, 2001; Ahmed and Garnett, 2011).

Bangladesh has two to three million hectares of land that is suitable for rice–fish based IAA production (ADB, 2005; Dey and Prein, 2006; Ahmed and Garnett, 2010; Dey et al., 2013). Recent estimates indicate that approximately 4.27 million households in the country (approximately 20% of all rural households) own a homestead pond that is suitable for IAA based fish production (Belton and Azad, 2012). However, due to small farm sizes and low levels of investment in relevant social, economic, and policy dimensions, this potential is not being fulfilled. According to another recent estimate only about 180,000 ha are currently under rice–fish based IAA production, well below the nation's potential. This raises the questions of whether the adoption and impacts of rice–fish based IAA systems are being adequately examined or not (Dey et al., 2013). Over many years Bangladesh's agricultural research and extension system, international organisations like WF, various domestic NGOs, private companies, and rural entrepreneurs have all contributed to extensive research and extension services in a participatory manner to achieve the country's IAA potential. One such initiative of WF was the AFP[21], a food security oriented effort to diversify rural livelihood options for resource-poor, marginalized *adivasi* (indigenous) communities

---

20 IAA is based on the concept of integrated resource management, utilizing synergies among subsystems that result in greater farm productivity. For detailed discussions of IAA related technologies see Edwards (1998), Prein (2002), and Pant et al. (2005).
21 The terms *adivasi*, indigenous, ethnic minority and tribal are used interchangeably in this study. In Bangladesh *adivasi* communities are typically the most marginalized and extremely poor segments of society; live in densely populated border areas; face dispossession and eviction from their ancestral lands; are often excluded from social

in northern and northwestern Bangladesh. Through participatory processes the project set out to devise and disseminate IAA technologies and related enterprise options (using a value chain approach) to match the existing physical and human asset bases, and the social and economic contexts and aspirations needs, resources, and capabilities of *adivasi* households (Pant et al., 2014). Thus participation in IAA value chains by indigenous households was not random.

Based on three rounds of panel data (2007, 2009, and 2012) from indigenous households, this study assessed the impacts of IAA value chain participation on the welfare of indigenous households in areas where the AFP operates. We also examined the distributional impacts of IAA value chain participation across different groups of value chain actors by disaggregating production activity participants from up and downstream value chain participants. Given the potential importance of IAA systems in Bangladesh this research mostly focused on rice–fish based IAA and the biophysical and technical feasibility aspects rather than socio-economic aspects. There have been previous socio-economic research efforts on IAA in Bangladesh (Ahmed and Garnett, 2011; Ahmed et al., 2011; Jahan and Pemsl, 2011; Dey et al., 2013) and elsewhere (Prein, 2002; Pant et al., 2005; Dey et al., 2010), but all of these efforts used cross-sectional data that makes it very difficult to control for unobserved heterogeneity and endogeneity.

Using a large and unique three-wave panel dataset and different panel analysis methods such as Fixed-Effects (FE), RE, and bias corrected FE models (Heckit panel), and control function approaches we hope to contribute to relevant literature in at least three ways. First, is to identify the casual effects of IAA value chain participation on household welfare in extreme poverty settings with due consideration for both observed and unobserved heterogeneity and endogeneity of IAA value chain participation. Second, is the consideration of backward and forward linkages along IAA value chains for a more comprehensive impact assessment that documents the evidence of heterogeneous treatment effects of IAA value chain participation. Third, is that this study took into account the impacts of the dynamics of IAA value chain participation over time, which is not possible using cross-sectional data and is seldom considered in many panel data based impact evaluation studies. This appears to be the first impact assessment of IAA technologies using a long-term panel dataset. The results of the analyses presented in this chapter provide valuable insights for other developing countries

---

safety net programs; are frequently trapped in poverty; and a significant proportion of them live below the absolute national poverty line (Pant et al., 2014).

with similar agro-ecological, socioeconomic, and institutional settings for efforts to address extreme poverty and marginality problems through IAA systems.

The rest of the chapter is organized as follows. Section 4.2 briefly describes the impact pathways framework of IAA value chain participation. Section 4.3 presents an overview of previous findings on impact of IAA and related technologies, including technology development and diffusion in Bangladesh. This is followed by a description of data and descriptive statistics in section 4.3 and of the empirical approach in section 4.4. In section 4.5 we present the results and discussion, and in section 4.6 we conclude with the highlights of the key findings and policy implications.

4.2 Framework of the Study

The framework of the study portrayed in Figure 4.1 reveals the pathways and linkages between IAA value chain participation and possible outcomes. It is very important to better understand the linkages between participation in IAA value chains and possible outcomes such as household income, changes to food security and dietary nutrition, and other welfare implications, which are not well understood, especially in marginalized poverty settings in Asia and particularly in Bangladesh. The overall research hypothesis is that participation in IAA value chains improves farm and non-farm productivity and efficiency by using social, natural, and human capital as well as reallocating resources, leading to increased income and food security and dietary nutrition among IAA value chain actors, and thus increased the overall wellbeing of IAA value chain participants.

Figure 4.1: Conceptual framework showing the welfare effect pathways of IAA value chain participation [Adapted after modification from von Braun (1988) and Dey et al. (2010)]

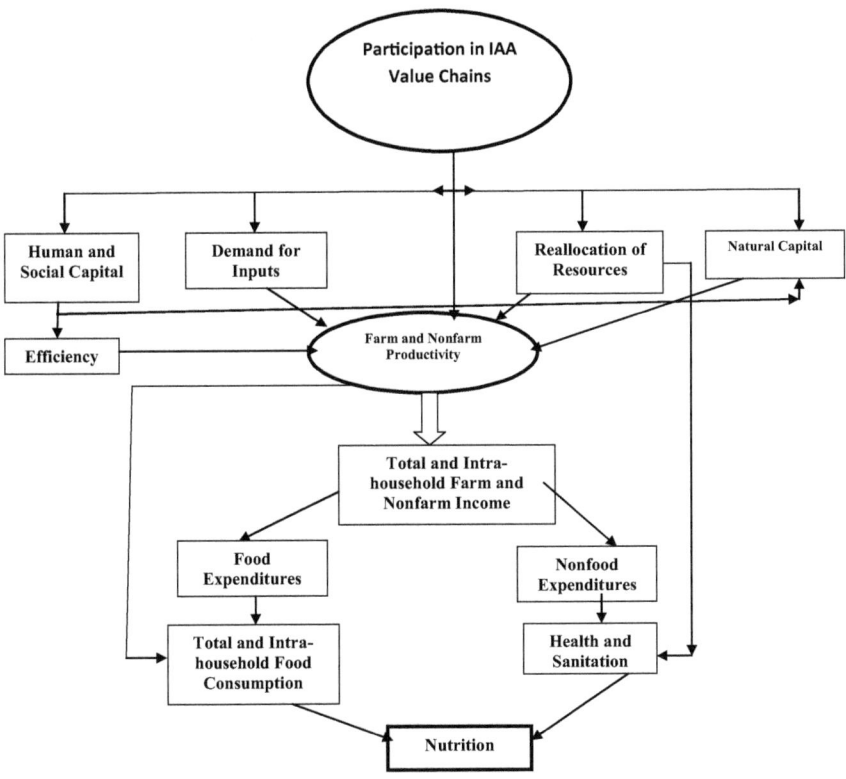

4.3 Literature Review

IAA systems originated in China as a means of increasing smallholder farmer productivity through integrated resource management (Brummett and Noble, 1995; Edwards, 2003). IAA utilizes synergies among subsystems through the flow of resources from one component of the system to another (Little and Muir, 1987; Edwards, 1993, 1998; Lightfoot et al., 1993; Brummett and Noble, 1995; Prein, 2002; Pant et al., 2005; Sugunan et al., 2006). Based on cross-sectional data from Malawi, Dey et al. (2010) found that adoption of IAA can improve farm productivity and efficiency. That study also found that IAA adoption helps increase total factor productivity, farm income, and returns on family labour. Brummett (1999) concluded that IAA offers the possibility of fostering the improvement of rural smallholder farmer livelihoods in Africa. Other studies have also found that IAA systems improve efficiency, productivity and ultimately farmer income (Karim, 2006; Dey et al., 2007). Based on an ex-ante analysis using a 'Trade-off Analysis for Multi-Dimensional Impact Assessment' (TOA-MD) model, Tran et al. (2013) showed that IAA systems could have positive impacts on poverty, household income, and food security among rural farming communities in Malawi. Using similar ex-ante evaluation methods Jahan et al. (2013) found more or less similar results in Bangladesh. Based on two-wave panel survey data Jahan and Pemsl (2011) claimed that IAA had significant positive impacts on profitability, productivity, and efficiency. They also found a positive effect on human and social capital, increased returns on family labour, better and more frequent access and higher level of engagement with local government, non-governmental and community-based organisations. Evidence from the Mekong delta in Vietnam indicates that the main incentives for adopting IAA farming there are positive income and consumption effects, and the reduction of negative environmental impacts. That study also found that IAA adoption increased agricultural diversification and cropping intensity (Nhan et al., 2007). Phong et al. (2010) found similar results from the same region. Research from Kenya has shown that IAA is a potential technique for contributing to household food security and livelihood improvements (Kipkemboi et al., 2007). Based on a review of related efforts in Asia, Prein (2002) concluded that IAA has considerable potential to improve the livelihoods of rural smallholder farmers.

Based on an extensive cross-sectional survey from Bangladesh Dey et al. (2013) found variable impacts of rice–fish based IAA participation. They asserted that innovation in rice–fish based IAA systems in the country is being driven primarily by households and community initiatives, and that there is need for more supportive government policies and investment to accelerate innovation. Another study on rice–fish based IAA in Bangladesh found that integration resulted in greater net benefits to farmers of 60% to 80% depending on the season and that mean rice yields increased by 10 to 12% (Gupta et al., 1996).

Based on research in Africa and Asia, Prein and Ahmed (2000) claimed that IAA systems can improve dietary nutrition and food security both directly and indirectly, and they recommended that much greater research efforts be made to achieve these benefits. Ruddle (1982) reported that rice–fish based IAA has great potential for supplying low-cost animal protein in the tropics and for increasing the income of smallholder farmers. Similarly, Kipkemboi et al. (2007) showed that pond-based IAA in Kenya has the potential to contribute to household food security and to improve livelihoods, but that biophysical variations should take into account to fulfil potential and reduce risks. Edwards (2000) suggested that with adequate support, aquaculture could contribute significantly to rural development in Asia where this kind of practice is traditional, as well as in countries where it is neither a traditional nor widespread practice such as in Africa.

Most of these examples of IAA research have focused primarily on biological and technical issues. Research on the socioeconomic impacts of IAA is very limited. One of the central methodological challenges to impact evaluation studies is addressing the attribution issue, and to establish causality between initiatives or technology and outcomes (Imbens and Wooldrige, 2009). Other interrelated research challenges include: i) establishing counterfactual positions; ii) addressing time lags; and iii) determining what level impact assessments should be performed at (macro versus micro) (Alston and Pardey, 2001; Salter and Martin, 2001). Most studies on technology adoption impacts in general, and on IAA adoption impacts in particular, are based on cross-sectional data, making it is difficult or even impossible to account for those problems. The use of long-term panel data in this study addresses some of these problems in determining the impacts of IAA value chain participation on household welfare.

4.4 Data and Descriptive Statistics

Data used in this chapter are based on three wave panel that is discussed in Chapter 1, section 1.5.2, and shows in Table 1.1.

4.4.1 Who Participated in Integrated Aquaculture-Agriculture Value Chains in Bangladesh?

Descriptive statistics of the socio-economic explanatory variables used in this analysis are shown in Table 4.1 by IAA participation category. Households participating in IAA value chains were, on average, headed by younger and more educated farmers. In addition, IAA value chain participators had larger families. This is consistent with the higher labour requirements of IAA value chain activities relative to rice monoculture. This implies that family labour has an important role in IAA participation and possibly indicates that subsistence pressure is part of the IAA value chain participation decision-making process. There were proportionately more male-headed households among participants than in the non-participant group.

Evaluation of changes in household IAA value chain participation status over the study period (2007–2012) showed that the probability of continued IAA value chain participation was lower than the probability of discontinuing participation and non-participation. Only 41% of households who participated in IAA value chains continued to participate in the subsequent period, while 38% of households dis-participate in subsequent periods and the rate of discontinued participation is quite high among the up and down steam IAA vale chain participants relative to production process participants. 'Dis-participators' differed from both 'non-participators' and continuous 'participators' in terms of almost all characteristics reported in Table 4.1.

Table 4.1: Descriptive statistics (mean and standard deviation) of Integrated Aquaculture-Agriculture participation explanatory variables in Bangladesh

Variables	Definition and measurement	2007		2009		2012		
		Non-participator	Participator	Non-participator	Participator	Non-participator	Dis-participator	Participator
Gender of HH head	Dummy (1 if Household head is male, 0 otherwise)	0.90 (0.30)	0.95 (0.21)	0.91 (0.28)	0.95 (0.22)	0.88 (0.33)	0.92 (0.28)	0.93 (0.26)
Age of HH head	Continuous (Age of household head in years)	46.1 (11.7)	43.7 (12.2)	47.9 (12.2)	46.0 (13.4)	50.1 (12.7)	47.9 (13.7)	48.3 (12.2)
HH size	Continuous (Total number of household members)	4.42 (1.51)	4.53 (1.62)	4.40 (1.61)	4.61 (1.58)	4.54 (1.69)	4.44 (1.54)	4.80 (1.57)
Farm size	Continuous (Total land area in decimals)	103.6 (107.3)	102.9 (122.9)	125.2 (187.7)	108.0 (117.8)	115.7 (172.7)	79.5 (117.1)	119.6 (123.5)
Non-farm income	Continuous (Per year in BDT)	21650.9 (16983.5)	21551.08 (15664.5)	30944.82 (28988.8)	28453.48 (24363.9)	28343.01 (44566.2)	32931.94 (50380.2)	36867.74 (54930.9)
CBO membership	Dummy (1 if household head is a member of a CBO, 0 otherwise)	0.40 (0.49)	0.98 (0.14)	0.58 (0.50)	1.00 (0.04)	0.18 (0.39)	0.13 (0.34)	0.24 (0.43)
Access to extension services	Dummy (1 if household had access to extension services)	0.94 (0.24)	0.93 (0.25)	0.93 (0.25)	0.95 (0.22)	0.41 (0.49)	0.38 (0.49)	0.52 (0.50)
Irrigation	Dummy (1 if irrigated crop land last year, 0 otherwise)	0.70 (0.46)	0.61 (0.49)	0.66 (0.47)	0.65 (0.48)	0.60 (0.49)	0.42 (0.49)	0.60 (0.49)
Access to credit	Dummy (1 if able to access credit, 0 otherwise)	0.92 (0.27)	0.91 (0.29)	0.78 (0.41)	0.86 (0.34)	0.79 (0.41)	0.71 (0.45)	0.75 (0.44)
Access to market information	Dummy (1 if agricultural market information available, 0 otherwise)	0.82 (0.39)	0.83 (0.38)	0.62 (0.49)	0.69 (0.46)	0.55 (0.50)	0.47 (0.50)	0.69 (0.46)
Marital status of HH head	Dummy (1 if the household head is married, 0 otherwise)	0.86 (0.34)	0.93 (0.25)	0.92 (0.27)	0.93 (0.26)	0.92 (0.28)	0.92 (0.28)	0.92 (0.27)
Main occupation of HH head	Dummy (1 if main occupation of household head is agriculture, 0 otherwise)	0.38 (0.49)	0.37 (0.48)	0.46 (0.50)	0.39 (0.49)	0.48 (0.50)	0.37 (0.48)	0.47 (0.50)
Education of HH head	Continuous (Number of years that household head attended school)	3.15 (3.93)	3.24 (3.83)	3.43 (3.83)	4.06 (4.08)	3.40 (3.80)	2.50 (3.43)	4.57 (3.99)
Number and (percentage) of observations		147 (22.5)	510 (77.5)	148 (22.5)	509 (77.5)	121 (21.2)	216 (37.8)	234 (41.1)

4.4.2 Relationships between Integrated Aquaculture-Agriculture Value Chain Participation Dynamics and Household Welfare

The IAA value chain participator groups are distinguishable in terms of welfare based on total income from land, other assets, off-farm income, and non-farm income. In the baseline (2007) incomes among IAA value chain 'participators' were lower than 'non-participators,' but in the subsequent survey year 'participator' incomes were significantly higher (Figure 4.2). All monetary values for 2007 and 2008 were deflated to 2012, in order to account for inflation and make the income values from the different survey rounds more comparable. To adjust monetary values we used the annual national consumer price index from the Bangladesh Bureau of Statistics (BER, 2012). Table A.4.1 in the Appendix provides details on the welfare outcomes by year and participators category. Income and the consumption frequency of food items such as fish, meat, pulses, fruits, and vegetables also increased among IAA value chain 'participators' relative to 'non-participators.' The results also indicate welfare implications of dis-participation relative to continued participation or non-participation. Although the outcome difference between 'dis-participators' and 'non-participators' was not significant, the difference between 'dis-participators' and continued 'participators' was significant. This suggests that dis-participation from IAA value chain activities may have negative welfare effects relative to continued participation.

Figure 4.2: Household income by integrated aquaculture-agriculture value chain participation categories in Bangladesh

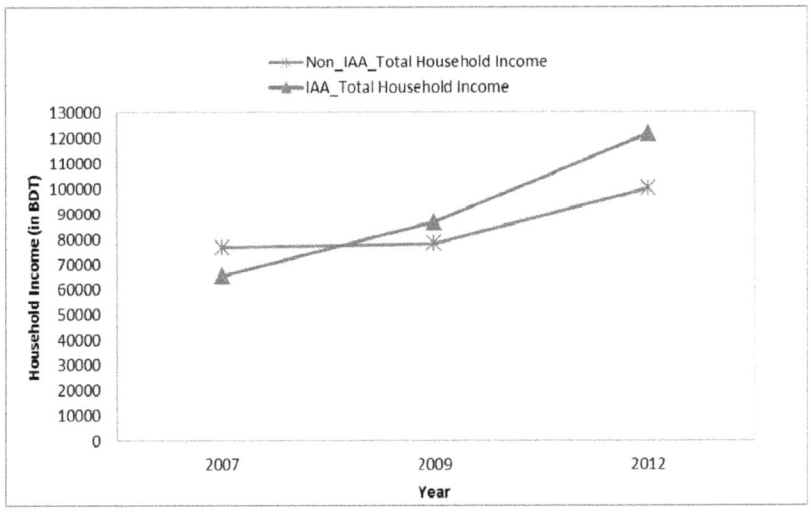

4.4.3 Distributional Impacts of Integrated Aquaculture-Agriculture Value Chain Participation

The welfare effects of IAA value chain participation may not be the same for all actors irrespective of their participation status. To test for income variability among participation groups, we disaggregated the IAA value chain participants into two groups: production activity participants and up and down stream (non-production) activity participants. The IAA production activity group was comparatively better off than up and downstream IAA value chain participants. Figure 4.3 presents the welfare outcomes by group and year. Welfare gains among IAA value chain participants increased over time for both groups, but the rate for production activity participants was much higher than for up and down stream participants. The details of the outcome differences between the two major groups are shown in Table A.4.2 in the Appendix. Those details also show that IAA production 'participators' had significantly greater total income than 'non-participators.' In order to be able to infer whether income differences were due to IAA value chain participation or other factors we applied a rigorous analytical model to identify if mean welfare outcomes were due to IAA value chain participation or not after controlling for confounding factors.

Figure 4.3: Distribution of income effects among integrated aquaculture-agriculture value chain participation categories in Bangladesh

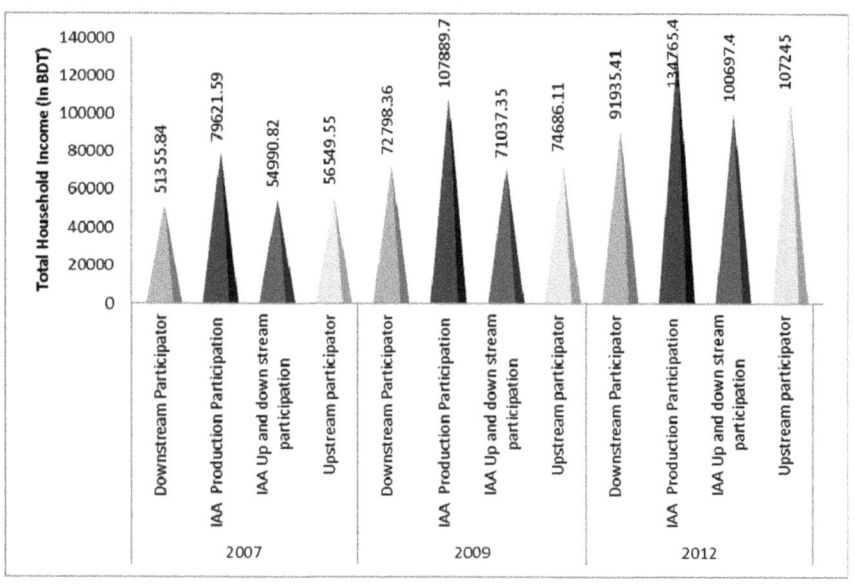

4.5 Estimation Issues and Strategy

The decision of whether or not to participate in IAA value chains is not random, thus the outcome of IAA value chain 'participators' and 'non-participators' are not directly comparable. This presents some challenges to estimating the welfare functions, particularly regarding how the unobserved heterogeneity and potential endogeneity of some variables are addressed in the models. In this section we discuss the estimated models and how these issues were treated.

In any given year the decision by a household to participate in IAA value chains or not will be determined by its expected utility or the relative costs and benefits associated with either option. Participation in IAA value chains was expected to have important positive impacts on household welfare through direct and indirect pathways that were discussed in the context of the conceptual framework. The indicators of household welfare outcomes for this analysis were annual household income and the consumption frequency of different food items measured at the household level. Another suitable welfare indicator, asset holdings (the value of household assets), was not an option because these data were not collected during the first two survey efforts. The welfare equation is simple and relatively straightforward. We defined welfare (income and consumption) (Y_{it}) as a function of IAA value chain participation ($IAAp_{it}$), IAA value chain dis-participation ($IAAd_{it}$), and a vector of relevant covariates (Z_{it}), that may include both time-variant and time-invariant factors.

The generic specification for impact evaluation is expressed as:

$$Y_{it} = \alpha + Z_{it}\beta_1 + IAAp_{it}\beta_2 + IAAd_{it}\beta_3 + T_t\gamma + c_i + \varepsilon_{it} \qquad (1)$$

In this impact model the dynamics of participation are shown by analysing whether the gains from IAA value chain participation persist over time, and what the economic impacts of dis-participation from IAA value chain. The coefficients (β_2 and β_3) of the two participation status dummy variables in this model represent the impacts of being in the particular category on welfare as compared to non-participation in IAA value chains, which was the reference category. β_2 indicates whether the range of welfare outcomes among IAA 'participators' and 'non-participators' increases or decreases over time for households that remain in their particular category. β_3 is the effect of dis-participation in comparison to non-participation in IAA value chains. The difference between coefficients ($\beta_2-\beta_3$) provides the welfare effects of dis-participation in comparison to IAA value chain participation. A dummy variable representing the year data were collected (T_t) was used to control for time fixed effects, c_i is an individual-specific effect, and ε_{it} is an idiosyncratic error term. Explanatory variables included in Z that are likely to affect household welfare are based on an extensive literature review on technology, or innovation, or high value chain participation impact studies, that are (Table 4.1) discussed in section 4.4.1.

At the beginning it was assumed that IAA value chain participation is exogenous (i.e. the decision to participate in IAA value chains is independent of material outcome) and there are presumably no factors that simultaneously affect IAA value chain participation and household welfare. We used a Pooled Ordinary Least Squares (POLS) estimator to estimate Equation 2.

$$Y_{it} = \alpha + Z_{it}\beta_1 + IAAp_{it}\beta_2 + IAAd_{it}\beta_3 + T_t\gamma + \varepsilon_{it} \qquad (2)$$

The POLS estimator ignores the panel structure of the data and simply estimates the coefficients by using OLS regression based on the assumption that ε_{it}, the idiosyncratic error term, is uncorrelated with the explanatory variables in Equation 2. Equation 2 was estimated in three different ways; first, as it is stated above, second, by using only IAA value chain participation ($IAAp_{it}$) and other covariates in Equation 2 (i.e. $Y_{it} = \alpha + Z_{it}\beta_1 + IAAp_{it}\beta_2 + T_t\gamma + \varepsilon_{it}$), third, by using IAA value chain dis-participation ($IAAd_{it}$) and other covariates in Equation 2 (i.e. $Y_{it} = \alpha + Z_{it}\beta_1 + IAAd_{it}\beta_3 + T_t\gamma + \varepsilon_{it}$).

It is very unlikely, however, that IAA value chain participation is exogenous. IAA value chain participation is a decision variable and hence may be correlated with the error term in the welfare equations. It may also result from unobserved heterogeneity between IAA value chain 'participators' and 'non-participators.' Such heterogeneity is very likely, as households self-select the IAA value chain participation category they belong to. Households that self-select IAA value chain participation may do so on the basis of unobservable characteristics and that also may determine the household welfare (Heckman and Hotz, 1989). Households with greater resources, skills, capabilities, and motivation (which are all also likely to affect household welfare) may decide to participate in IAA value chains, while those that do not have or that have fewer resources may not participate in IAA value chains and vice versa. If this is the case the impact of IAA value chain participation estimated with Equation 1 will be either over or underestimated.

Panel data models that allow IAA value chain participation decisions to be correlated with unobservable effects on outcome variables control this problem (Heckman and Holtz, 1989; Berhane and Gardebroek, 2011). We used three such empirical approaches to exploit the panel nature of the data: the Standard FE model, the Heckit panel model, and a control function approach. An important issue in estimating panel models is how to deal with the unobserved heterogeneity effect, ci. Following a strict exogeneity assumption (i.e. the time invariant unobserved heterogeneity, ci, is not correlated to any of the other covariates) then

$v_{it} = c_i + \varepsilon_{it}$ can be considered as a composite error and the following equation can be estimated through a RE model. The RE model was estimated in three different ways like the POLS model.

$$Y_{it} = \alpha + Z_{it}\beta_1 + IAAp_{it}\beta_2 + IAAd_{it}\beta_3 + T_t\gamma + v_{it} \qquad (3)$$

However, the strict exogeneity assumption is very strong and it is very unlikely that the unobserved heterogeneity will be orthogonal and uncorrelated to the other covariates (Bezu et al., 2014). If this is the case and it is not controlled for, this could lead to selection bias in the estimated welfare effects of IAA value chain participation. Literature suggests that a common and straightforward way to control the selection bias problem is to use a household FE estimator (Wooldridge, 2010b; Greene, 2003). Recent empirical research efforts frequently use an FE estimator to control for the selection bias problem (Crost et al., 2007; Jorgenson and Birkholz, 2010; Berhane and Gardebroek, 2011; Kouser and Qaim, 2011; Kathage and Qaim, 2012; Bezu et al., 2014; Muriithi and Matz, 2015).

This Standard FE model allows for individual heterogeneity, ci, to be correlated with the vector of explanatory variables, Z_{it}. The Standard FE estimator provides a consistent estimate of welfare effects by differencing out all time-invariant unobserved heterogeneity effects (Wooldridge, 2010b). We estimated the welfare outcome Equation 1 by using a FE model because these are linear models. Some of the outcome variables are censored at zero or count variables. For such outcome variables (such as consumption frequency) in the welfare equation we estimated the equation by using a Poisson estimator, otherwise the linear specification will lead to a biased estimate (Wooldridge, 2010b). We also used a Hausman test to compare the FE and RE model results and to detect unobserved heterogeneity, although it is neither necessary nor a sufficient condition to test (Snijders, 2005; Greene, 2003). The results from both models are reported, but an interpretation is only given for the FE estimate.[22]

---

22 Within-group variability with respect to the treatment variable (in this case IAAp and IAAd) is necessary in order to estimate an efficient FE model (Kikulwe et al., 2014). Thus, there needs to be a sufficient number of households that participate in IAA value chains or that discontinued participation in the first year of the survey, but not in another year. Such variability is present in the data, especially between the survey wave one (2007) and three (2012), and between years two (2009) and three (2012), because in the third wave a large number of IAA value chain 'participants' became 'dis-participators.' Variability in IAA participation in this case only comes from dis-participation in IAA value chain. Thus the FE model was estimated by using only an IAA participation dummy and other covariates. Otherwise different estimates for

As already discussed, IAA value chain participation from Equation 1 may be correlated with the error term. We also used a framework similar to Heckman's two-stage model with panel settings to control for possible endogeneity of the selection of participation in IAA value chains and checked the robustness of the above results with a FE estimator. The first step involves estimating the IAA participation selection equation using a pooled probit model for different T by including the exclusion restriction variable and then computed the T inverse Mills' ratios (λ_{1it} for participation). In the second step the Mills ratios were plugged into the welfare outcome equation to control for possible self-selection into IAA value chain participation, which was then estimated using a standard household FE model by excluding the exclusion restriction variable. The welfare outcome Equation 1 was amended as follows:

$$Y_{it} = \alpha + X_{it}\beta_1 + IAAp_{it}\beta_2 + \lambda_{1it} + T_t\gamma + c_i + \varepsilon_{it} \qquad (4)$$

where X_{it} contains Z_{it}, but one variable less, that affect IAA value chain participation, but not household welfare, which is what the identification of the causal effect is dependent on.

Furthermore a control function approach was used to control for possible endogeneity of selection in IAA value chain participation and to check the robustness of the above results with a different estimator. This approach also involves two steps; the first step it involves estimating the reduced form (like the selection model) of the IAA participation model by using a RE probit model by including the exclusion restrictions and then computing the generalized residuals (δ_{1it} for participation). In the second step the generalized residual was included in the welfare outcome (structural) equation to control for possible endogenous selection of IAA value chain participation, which is then estimated using a standard household FE model by removing the exclusion restriction variable. A significance test on the coefficient (δ_1) of the residuals tests for endogeneity of the IAA value chain participation (Bezu et al., 2014). The welfare outcome Equation 1 was amended as follows:

$$Y_{it} = \alpha + X_{it}\beta_1 + IAAp_{it}\beta_2 + \delta_{1it} + T_t\gamma + c_i + \varepsilon_{it} \qquad (5)$$

In the Heckit model and control function approach 'CBO membership status' is used as an exclusion restriction. This variable may influence household participation in IAA value chains. This variable is considered a viable exclusion restriction

---

IAA participation and dis- participation will give similar results because the change in participation status is the mirror image of change in dis-participation. Thus the dis-participation effect on welfare was based on the POLS and RE models because they do not need within variability.

because CBO membership is expected to represent social capital at the individual and village levels because these are voluntary membership organisations and among indigenous people participation in this type of organisation is very high and they have high degree of social cohesion (Pant et al., 2014). Thus this variable was included in the participation equation to satisfy the exclusion restriction in the above two models, which are not included in the welfare outcome equations. These variables were not expected to affect the welfare outcome equations directly after controlling for IAA participation.

4.6 Results

4.6.1 Integrated Aquaculture-Agriculture Value Chain Participation Dynamics and Household Welfare

4.6.1.1 Naive Pooled Ordinary Least Squares and Random Effects model results

The estimation results from different specifications (separate regression analyses for participation and dis-participation as well as both in the same equation) of the POLS and RE models are shown in Table 4.2. The results of both models indicate that overall, IAA participation had a positive and significant relationship with household income. On the other hand dis-participation from IAA was negatively associated with household income. The magnitude of the coefficient shows that IAA participation increased annual income by almost 5,500 Bangladeshi Taka (BDT). Combining participation and dis-participation in same equation (columns 3 and 6) had similar results. Thus IAA participation was positively associated with household welfare and dis-participation was negatively associates with household welfare. The latter was expected and it indicates that the benefits of IAA participation do not continue to accrue after discontinuing participation, suggesting that the decision to discontinue participation is not due to the economic superiority of this option. However, the results from the POLS and RE models should be interpreted cautiously, as the decision to participate in IAA value chains is not exogenous and the specification needs to be modified to address this issue.

Table 4.2: Pooled Ordinary Least Squares and Random Effects model results

Variables	Total Household Income†					
	Pooled OLS			RE GLS regression		
	Coef. (Robust Std. Err.)	Coef. (Robust Std. Err.)	Coef. (Robust Std. Err.)	Coef. (Robust Std. Err.)	Coef. (Robust Std. Err.)	Coef. (Robust Std. Err.)
	1	2	3	4	5	6
Participation in IAA value chain	5.4** (2.4)		3.4 (2.7)	5.7** (2.8)		3.3 (2.8)
Dis-participation from IAA value chain		−10.8*** (4.1)	−8.7*** (4.4)		−11.0*** (5.1)	−8.8* (5.3)
Year 2009	14.5*** (2.6)	14.5*** (2.6)	14.5*** (2.6)	14.4*** (2.0)	14.5*** (2.0)	14.4*** (2.0)
Year 2012	42.2*** (3.2)	44.2*** (3.5)	44.5*** (3.5)	42.0*** (4.0)	44.0*** (4.3)	44.3*** (4.3)
Age of HH head	0.1 (0.1)	0.1 (0.1)	0.1 (0.1)	0.1 (0.1)	0.1 (0.1)	0.1 (0.1)
Education of HH head	1.4*** (0.3)	1.3*** (0.3)	1.3*** (0.3)	1.4*** (0.3)	1.3*** (0.3)	1.3*** (0.3)
Total HH size	8.3*** (0.7)	8.4*** (0.7)	8.3*** (0.7)	8.4*** (0.8)	8.4*** (0.8)	8.4*** (0.8)
Farm size	0.2*** (0.0)	0.2*** (0.0)	0.2*** (0.0)	0.2*** (0.0)	0.2*** (0.0)	0.2*** (0.0)
Access to extension	4.1 (3.2)	3.9 (3.2)	3.8 (3.2)	4.1 (3.8)	3.9 (3.8)	3.8 (3.8)
Irrigation	7.4*** (2.4)	6.9*** (2.4)	7.1*** (2.4)	7.2*** (2.6)	6.8*** (2.6)	6.9*** (2.6)
Access to credit	8.0*** (3.0)	8.0*** (3.0)	8.0*** (3.0)	7.6*** (3.3)	7.6*** (3.3)	7.6*** (3.3)
Access to market information	2.9 (2.5)	2.9 (2.5)	2.7 (2.5)	2.7 (2.6)	2.7 (2.6)	2.5 (2.6)
Constant	−18.4*** (6.6)	−13.3*** (6.5)	−15.7** (6.7)	−18.4*** (7.3)	−13.1 (7.4)	−15.5*** (7.5)
R^2	0.36	0.36	0.36	0.36	0.36	0.36
Number of observations	1885	1885	1885	1885	1885	1885
Number of groups				657	657	657
Rho				0.1	0.1	0.1
F (11, 1873)/Wald chi² (11)	94.67***	94.99***	87.24***	509.71***	507.3***	510***

Notes: † The total household income (dependent) variable is in thousands; * significance at 10%; ** significance at 5%; *** significance at 1%.

4.6.1.2 Standard Fixed Effects model results

We applied the FE models to reveal the relationship between IAA participation[23] and household welfare. The selected indicators for welfare outcomes are annual household income and household consumption frequency of selected food items. Household consumption frequency was computed by counting the number of times that a household consumed a particular food item over the course of a day, or week, or month depending on the item. Household income includes income from crops, livestock, fisheries, non-farm activities, and off-farm activities.

Table 4.3 presents the results of the FE models of the impacts of IAA participation on household income and consumption respectively. IAA value chain participation had a large positive and significant association with household income. The FE estimates show that, controlling for other factors, IAA participation was associated with an increase of approximately 13,500 BDT in household income and the time FE model results show that this effect has increased over time. This is notable given the fact that sample households are indigenous, which are one of the mostly marginalized and extremely poor socio-ethnic groups in Bangladesh whose members typically own small properties, and sustainable intensification using IAA is a potential option for increasing food production. We also ran a Hausman test and the results reject the RE model results in favour of the FE model results. Similarly, the FE model results for the relationship between IAA participation and household consumption frequency show that IAA participation was significantly and positively associated with increased fish, meat, egg, pulse and vegetable consumption frequency and that the consumption of rice and fish continued to increase over the course of the year. In the consumption equation we also controlled for income effects by using lagged household income[24] to compare the income and participation effects. The results indicate that IAA value chain participation effects are greater than income effects.

---

23 Here we only used IAA participation in the FE and subsequent models because of the variability of IAA participation resulting from dis-partiticaption. Thus changes in the IAA participation are the mirror image of changes in the dis-participation.

24 Although theoretically (e.g. according to agricultural household models) consumption is a function of income and other covariates, but empirically income is endogenous to consumption and in this case current income is highly correlated with IAA participation. Here we used lagged income instead of current income to avoid the endogeneity problem (e.g. Asfaw et al. 2012 also used lagged income instead of current income for estimating welfare impacts of technology adoption).

Table 4.3: Fixed Effects model results of the relationship between IAA value chain participation and household income and consumption frequency of selected foods in Bangladesh

Variables	Total Household Income†	Household Consumption Frequency					
		Rice	Fish	Meat	Egg	Pulses	Vegetables
	Coef. (Robust Std. Err.)	Coef. (Robust Std. Err.)	Coef. (Robust Std. Err.)	Coef. (Robust Std. Err.)	Coef. (Robust Std. Err.)	Coef. (Robust Std. Err.)	Coef. (Robust Std. Err.)
Participation in IAA value chains	13.5*** (5.3)	0.02 (0.02)	0.37*** (0.07)	0.21* (0.12)	0.32** (0.14)	0.22*** (0.1)	0.10** (0.04)
Year 2009	14.8*** (2.0)	–	–	–	–	–	–
Year 2012	41.0*** (4.4)	0.04*** (0.01)	0.13*** (0.05)	0.13 (0.09)	−0.35*** (0.11)	−0.25*** (0.06)	0.04 (0.03)
Lagged household income		0.00** (0.00)	0.00 (0.00)	0.00 (0.00)	0.00 (0.00)	0.00 (0.00)	0.00** (0.00)
Age of HH head	0.4* (0.2)	0.00 (0.00)	0.00 (0.00)	0.00 (0.00)	0.00 (0.00)	0.00 (0.00)	0.00 (0.00)
Total HH size	8.7*** (1.8)	0.00 (0.01)	0.03* (0.02)	0.03 (0.03)	0.00 (0.04)	0.04 (0.03)	0.01 (0.01)
Farm size	0.1*** (0.0)	0.00 (0.00)	0.00 (0.00)	0.00** (0.00)	0.00* (0.00)	0.00 (0.00)	0.00 (0.00)
Access to extension services	2.7 (3.9)	−0.01 (0.02)	−0.09 (0.06)	−0.15 (0.10)	−0.18 (0.13)	0.08 (0.07)	−0.02 (0.04)
Irrigation	3.0 (3.3)	0.01 (0.02)	0.03 (0.07)	0.03 (0.11)	−0.07 (0.13)	0.14** (0.07)	−0.02 (0.04)
Access to credit	3.4 (4.0)	0.00 (0.02)	0.02 (0.07)	0.02 (0.10)	0.32** (0.14)	0.06 (0.08)	0.05 (0.04)
Access to market information	−0.9 (3.2)	−0.02 (0.02)	−0.16*** (0.05)	−0.17** (0.09)	−0.29*** (0.11)	−0.12** (0.06)	0.024 (0.03)
Constant	−13.3 (13.9)						
Rho	0.34						
R^2 overall	0.31						
Number of observations	1885	1142	1140	1060	898	1102	1142
Number of groups	657	571	570	530	449	551	571
F (10, 656)/Wald chi^2 (10)	20.08***	25.24***	68.03***	19.32**	45.97***	25.77***	22.45***

Notes: † The total household income (dependent) variable is in thousands; * significance at 10%; ** significance at 5%; *** significance at 1%.

4.6.1.3 Heckit panel and control function approach results

To check the robustness of the FE estimates and control for possible selection bias in IAA participation, we used a Heckman bias corrected FE model where the inverse Mills ratios from the first stage selection pooled probit are used to control for both individual-specific, time-variant observable and time-invariant unobservable characteristic related selection problems. We also used the control function approach to control for the possibility of endogenous selection of IAA participation. The generalized residuals from the first stage participation equations were included in the FE models to test and control for the endogeneity of IAA value chain participation. Both the Heckit FE and control function FE model results are reported in Table 4.4. Surprisingly the coefficient estimates of the inverse Mills ratios and generalized residuals for IAA value chain participation were not statistically significant, indicating that IAA value chain participation is not endogenous, which ameliorates concerns about the potential for endogenous selection bias with regard to IAA value chain participation.

Table 4.4 Heckit panel and control function model coefficients of the relationship between IAA participation and household income in Bangladesh

Variables	Household Income†	
	Control function Coef. (Robust Std. Err.)	Heckit Model Coef. (Robust Std. Err.)
Participation in IAA value chain	16.2*** (6.6)	14.0*** (5.7)
Year 2009	14.6*** (2.0)	14.9*** (2.0)
Year 2012	43.2*** (5.6)	40.2*** (4.8)
Inverse Mills ratio (IMR)		1.7 (5.1)
Generalized residual	−0.9 (1.3)	
Age of HH head	0.4 (0.2)	0.4 (0.2)
Total HH Size	8.8*** (1.8)	8.8*** (1.8)
Farm size	0.1*** (0.0)	0.1*** (0.0)
Access to extension services	2.7 (3.9)	2.8 (3.9)
Irrigation	2.9 (3.3)	2.9 (3.3)
Access to credit	3.3 (4.0)	3.4 (4.0)
Access to market information	−1.1 (3.2)	−0.7 (3.1)
Constant	−15.8 (14.9)	−14.5 (13.9)
Rho	0.34	0.34
R^2: overall	0.32	0.31
Number of observations	1876	1885
Number of groups	657	657
F (11, 656)	19.6***	18.6***

Notes: † The total household income (dependent) variable is in thousands of BDT; * significance at 10%; ** significance at 5%; *** significance at 1%.

The results reported in Table 4.4 indicate that IAA value chain participation is positively and significantly associated with household income, which is consistent with the earlier FE results. Interestingly, the coefficients are higher in magnitude compared to the FE results presented in Tables 4.3. Specifically, with IAA value chain participation household income increased by about 14000 to 16000 BDT depending on the model. Thus the positive income effect of IAA participation is robust under different specifications.

4.6.2 Who Benefits More From Integrated Aquaculture-Agriculture Value Chain Participation?

We calculated disaggregated results of the household income equation to compare comparatively wealthier households (that participated in IAA production related value chain activities, which consists of actors who require access to land) with poorer households (extremely poor households that participated in both up and down stream chain and some production related IAA activities, most of this group do not have access to land).

Table 4.5: *Comparison of the Fixed Effects model results of annual household income# for the IAA value chain actors by relative wealth in Bangladesh*

Variables	IAA Value Chain Participation Stage/Status			
	Up and down stream value chain activity actors††		Production value chain activity actors†	
	Coef.	Robust Std. Err.	Coef.	Robust Std. Err.
IAA value chain participation	0.07	0.08	0.32***	0.13
Year 2009	0.12***	0.03	0.20**	0.04
Year 2012	0.37***	0.05	0.44**	0.06
HH head age	0.00	0.00	0.00	0.00
Total HH size	0.10***	0.02	0.09***	0.02
Farm size	0.00***	0.00	0.00***	0.00
Access to irrigation	0.05	0.05	0.05	0.06
Access to credit	0.05	0.05	0.10	0.07
Access to market information	0.02	0.04	0.07	0.06
Access to extension services	0.05	0.05	0.09	0.07
Constant	10.14***	0.15	10.14***	0.21
Rho	0.33		0.38	
Number of observations	1253.00		1048.00	
Number of groups	441.00		364.00	
R^2 overall	0.3030		0.2424	

Notes: # The annual household income (dependent) variable is a logarithmic term; † actors that participated in production activities include: integrated rice–fish producers and integrated pond-fish producers (comparatively wealthier households); †† Upstream and downstream value chain actors include fingerling traders, fish traders, fishermen, cage cultivators, and community based fish and aquatic animal producers (comparatively poorer/landless households); * significance at 10%; ** significance at 5%; *** significance at 1%.

Table 4.5 presents the results from a separate FE model of income for comparatively wealthier and poorer households. The estimated participation in IAA value chain coefficients shows that participation in IAA value chains was positive for all households regardless of the participation activities, but coefficients for production related IAA actors were significantly and comparatively greater than up and downstream IAA value chain actors. An increase in IAA value chain participation was associated with an increase in household income of 32% for production related actors and only 7% for non-production IAA value chain activities that do not require land. Greater income effects among households involved in IAA production activities may reflect greater participation in IAA activities as these activities required land and higher capital investment than up and downstream activities. It seems that all IAA value chain actors do not have the same potential to capture benefits.

4.7 Conclusions

Bangladesh is one of the poorest and densely populated countries in the world. Rice and fish are the major staple foods and are highly associated with Bangladeshi culture. Rice alone accounts for 77% of agricultural land use, provides about 70% of direct human calorie intake, 66% of total daily dietary protein intake, and contributes more than 80% of the total food supply, but there are numerous constraints (e.g. negative environmental impacts of rice monoculture) affecting this sector that threaten sustainability (Bhuiyan et al., 2002; BRKB, 2010). Similarly, the fisheries sector also has a prominent role, contributing about 5% to GDP, 6% to export earnings, and 63% of dietary animal protein intake and there is a further potential for this sector to contribute more (Jahan and Pemsl, 2011; BER, 2014). In Bangladesh rice and fish demand is continuously increasing due to population growth (Chowdhury, 2009). Sustainably increasing agricultural productivity to enhance food security and dietary nutrition is a serious challenge in Bangladesh.

It has been argued that the adoption of agricultural technologies, or in this case IAA, increases food security and dietary nutrition through direct and indirect pathways. In Bangladesh, different national and international organisations (e.g. WF) and experimentation by farmers have developed locally appropriate IAA technologies and practices that have been promoted through both public and private sector initiatives such as the Development of Sustainable Aquaculture Project and the AFP that work to provide sustainable intensification options for smallholder farmers (Samsuzzaman, 2002; Karim, 2006; Jahan and Pemsl, 2011; Dey et al., 2013; Pant et al., 2014).

We investigated the relationships between IAA value chain participation dynamics among smallholder indigenous households and their economic wellbeing in order to contribute to the on-going debate of whether or not IAA is a sustainable intensification option that can contribute to poverty reduction and food security in developing countries, especially in Asia. We used a large three-year panel dataset collected from 2007 to 2012 to systematically address these effects. In addition to using the panel data and exploiting the possibilities associated with this type of data, this study was intended to contribute to relevant literature by examining the impacts of IAA value chain participation and dis-participation on two measures of household welfare: income and consumption. This also appears to be the first study of this type that explores the dynamic impacts of IAA participation by considering all value chain actors.

We estimated the welfare impacts of IAA participation dynamics using different models (e.g. POLS, RE, FE, Heckit, and control function models) under different assumptions to control for unobserved heterogeneity and endogenous selection of IAA value chain participation. We started with a naive POLS model that assumes that IAA value chain participation and dis-participation are exogenous. Subsequently, we controlled for unobserved heterogeneity and endogenous selection to validate the results. Additionally, we applied FE models to sample households disaggregated by value chain activity among production related actors (who participated in production related IAA value chain activities and that require land) and up and downstream value chain actors (extremely poor households, most of which do not have access to land), to explore the distribution of IAA value chain participation benefits across all IAA value chain actors.

The results are robust across specifications, thereby justifying our concerns about unobserved heterogeneity with respect to IAA value chain participation. We found consistent evidence of a positive relationship between IAA value chain participation and household income and the consumption of fish meat, egg, vegetables, and pulse beans. The results indicate that IAA value chain participation is associated with greater household income and this effect increased over time. Moreover, we found that IAA value chain participation was positively correlated with household income of both relatively poor (extremely poor households that participated in a variety of IAA value chain activities that did not require land) and wealthier households (that participated in IAA value chain production activities that required land), but that the benefits from IAA value chain participation were comparatively higher for the wealthier households. Considering participation dynamics, many of the former IAA value chain participants decided to discontinue participation, not based on the economic superiority of alternative options, but due to other barriers to IAA value chain participation.

Overall, we conclude that IAA value chain participation increases the welfare of poor and marginalized indigenous households in Bangladesh. The results show the importance of IAA value chain activities for poor smallholders and how IAA value chain participation may contribute to food security and poverty reduction among rural smallholders. Cost effective agricultural policies that help to create an enabling environment for sustainable technology adoption and continuation can significantly contribute to improved food security and poverty reduction in rural areas.

Further research using other alternative welfare indicators, such as an asset index that considers the quantity of assets and their monetary value would be helpful to better understand changes in the capital stocks of households. In addition, technology adoption scenarios like IAA value chain participation not only have direct impacts, but also may have indirect impacts (e.g. spill over effects), which are beyond the scope of this study. Thus future research that takes into account these broader economy-wide effects using appropriate economy-wide modelling approaches (e.g. Subramanian and Qaim, 2009) would provide a better understanding of IAA participation's broader impacts.

Chapter 5: Comparative Socio-environmental Impacts of Rice–Fish Based Integrated Aquaculture-Agriculture and Rice Monoculture in Bangladesh

5.1 Introduction

With a population of 155.8 million people, Bangladesh is one of the poorest and most densely populated (1203 persons per km^2) countries in the world (Ahmed et al, 2011; BER, 2014). Agriculture is the key economic sector, contributing 13% of its GDP, accounting for 48% of its total employment, and either directly or indirectly supporting the livelihoods of about 90% of its citizens (BER, 2014). Like in many other Asian countries rice and fish are staple foods in Bangladesh (Gupta et al., 1996; Dey et al., 2005, 2014; Kawarazuka and Béné, 2010). Rice production alone accounts for 76% of the country's cultivated area, 78% of the irrigated area, 52% of agricultural GDP, and 71% of daily caloric intake (BBS, 2012).

Efforts to intensify rice production have been implemented across Asia, including Bangladesh, through the rapid diffusion of GR technologies and practices (modern seed development, agro-chemical, and irrigation use), and associated complementary inputs to meet the rising demand for food (Hossain, 2010). Bangladesh has since been transformed from a net importer of rice to a nearly self-sufficient rice producer (Ahmed et al., 2014; BER, 2014). While the changes brought by the GR dramatically increased rice yields and transformed the lives of millions of rural, resource poor people across Asia (Hayami and Ruttan, 1985; Lipton and Longhurst, 1989; Hossain et al., 1990), its implementation has had unanticipated costs in terms of environmental unsustainability[25] (Brown, 1988; Redclift, 1989; Pimentel and Pimentel, 1990; Shiva, 1991; Rola and Pingali, 1993;

---

25 For greater detail on the negative consequences of GR practices and technologies see: Rola and Pingali (1993), Antle and Pingali (1994), Heong et al. (1995), and Huang et al. (2005) for impacts on public health; Way and Heong (1994) and Schoenly et al. (1996) for impacts on arthropod food webs; Moulton (1973), Cagauan and Arce (1992), Dashu et al. (1992), Gupta and Mazid (1993), Bottrell and Weil (1995), Halwart (1995), Abdullah et al. (1997), Halls et al. (1998), Shankar et al. (2004), and Klemick and Lichtenberg (2008) for impacts on aquaculture and aquatic food webs; and Shiva (1991), Ali et al. (1997), and Singh (2000) for impacts on soil fertility and genetic diversity.

Antle and Pingali, 1994; Singh, 2000; Klemick and Lichtenberg, 2008). This issue along with the increasing global population, the limitations of agricultural resources (e.g. land and water), and the effects of global climate change on crop production put unprecedented pressure on global food security (Xie et al., 2011a).

Therefore modern agricultural strategies need to be reconsidered (Bromley, 2010; Godfray et al, 2010), and recently such reconsideration has gained momentum, leading to discussions about sustainable agricultural intensification (Montpellier Panel, 2013). In Bangladesh the chemically intensive nature (e.g. synthetic chemical fertilisers, pesticides, and herbicides) of GR innovations is coupled with high cropping intensity, and rice monoculture with negative impacts on the environment, human health, and long-term food security (Alauddin and Tisdell, 1991; Rahman, 2005; Coelli et al., 2003). In addition, Bangladesh's rice production has already reached its maximum capacity in terms of available arable land, leaving very little or no potential for horizontal expansion (Hossain, 1988; Rahman, 2003a). The potential further conversion from local to modern rice varieties also seems to be limited, as this phenomenon is stabilising, particularly for varieties suited for irrigated production systems (Bera and Kelly, 1990; Hossain et. al., 2006).

Therefore, sustainable intensification of modern agriculture (especially rice production) to meet growing demand will be a challenge in Bangladesh and various researchers have suggested IAA as an alternative to simply trying to intensify rice monoculture production. Relative to rice monoculture systems, rice–fish based IAA systems are considered more sustainable intensification options because they maximise the benefits from increasingly scarce land and water resources by using relatively fewer chemical inputs, producing both dietary carbohydrate and protein resources, and having less negative impacts on biodiversity (Dela Cruz, 1994; Cagauan, 1995; Halwart, 1995, 1998; Frei and Becker, 2005a; Xie et al., 2011b).

So far research on rice–fish based IAA has been dominated by biological and technical issues and the general consensus is that it has considerable potential for more sustainable agricultural development. The benefits of rice–fish based IAA include: 1) that it is also an effective IPM strategy for some rice pests (Cagauan, 1995; Halwart, 1995, 1998; Halwart and Gupta, 2004; Dela Cruz, 1994; Berg, 2001), 2) that it improves agricultural diversification, intensification, productivity, and sustainability (Duong et al., 1998; Das et al., 2002; Gurung and Wagle, 2005; Halwart, 2006; Ahmed et al., 2007; Nhan et al., 2007), 3) that it improves soil fertility by generating nitrogen and phosphorus (Lightfoot et al., 1992; Giap et al., 2005; Dugan et al., 2006), 4) that it supports efforts to control malaria vector mosquitoes and water-borne diseases (Matteson, 2000), 5) that it can function as an important tool for climate change adaptation (through diversification) and the mitigation of environmental impacts (by using less environmentally hazardous inputs) (Lightfoot, 1990; Prein et al., 1998; Pullin, 1998; Rothuis et al., 1998b; Berg, 2002; Gooley and Gavine, 2003; Bosma et al., 2012).

Although there is huge potential for rice–fish based IAA systems in over 90% of the world's irrigated and rain fed rice production areas, it continues to be a marginal farming system and adoption has remained limited (Frei and Becker, 2005a; Halwart, 2006). In Bangladesh where rice–fish based IAA systems are particularly relevant because of the importance of both rice and fish as staple foods in Bangladeshi culture, its uptake also remains marginal (Nabi, 2008; Ahmed and Garnett, 2010; Ahmed et al., 2011; Dey et al., 2013). Bangladesh has about two to three million hectares of rice fields that are suitable for rice–fish based IAA systems (ADB, 2005; Dey and Prein, 2006; Ahmed and Garnett, 2010), but only about 0.18 million hectares have been dedicated to these efforts so far (Dey et al., 2013). There exists tremendous potential as well scope for expansion of rice–fish based IAA production areas (Wahab et al., 2008). Bangladesh's agricultural research and extension system and various organisations such as the BFRI, DOF), BRRI, BAU, and WF have all contributed to the promotion of rice–fish based IAA systems. However, despite these manifold efforts rice monoculture is still the dominant farming system in the country (Nabi, 2008; Ahmed and Garnett, 2010; Ahmed et al., 2011; Dey et al., 2013). This raises the question of whether or not the criteria and methods used to promote rice–fish based IAA farming systems in Bangladesh have been appropriate, or the rate of adoption has been adequately examined, and whether or not all socio-economic and environmental outcomes have been captured.[26]

In order to achieve the environmental, economic and social sustainability potential of rice–fish based IAA systems, farmers must decide whether or not to adopt them, which depends on the alternative options available to farmers and their perceptions about the costs and benefits of each option. Striving for utility maximisation, farmers are likely to choose technological options with the greatest expected potential for improving profits while minimising risk. The levels of knowledge and information available to individual farmers play crucial roles in their decision-making processes. Therefore, measuring farmer perceptions or awareness of the socio-environmental impacts of rice–fish based IAA system

---

26 A recent meta review by Dey et al. (2013) indicated that although the rice–fish based IAA farming systems have potential for more sustainable agriculture development and important implications for socio-economic, food security, and dietary nutrition improvement in Bangladesh, the social and economic research on the performance of rice–fish systems is relatively limited.

relative to rice monoculture and identification of socioeconomic determinants of such awareness could contribute to the design of more appropriate extension programmes in order to promote more sustainable agricultural technology in general, and rice–fish based IAA system in particular, among rural farmers in Bangladesh.

The main objective of the analyses presented in this chapter was to evaluate whether or not rice–fish based IAA farming systems provide sustainable alternatives to rice monoculture systems. To accomplish this objective this study followed four steps: i) a comparison of the relative environmental risks of inputs used at the farm level, ii) an extensive literature review to help interpret the significance of the study findings, iii) a comparison of farmer perceptions of the socio-environmental impacts associated with rice–fish based IAA and rice monoculture systems, and iv) an investigation of the factors affecting farmer perceptions about the socio-environmental impacts of rice monoculture systems.

Many rice–fish based IAA farmers also or formerly practiced monoculture rice production, whereas typical rice monoculture farmers were not familiar or even aware of rice–fish based IAA systems. We hypothesized that rice–fish based IAA farmers are more aware of the relatively positive socio-environmental impacts of rice–fish based IAA system compared to rice monoculture and the negative socio-environmental impacts of rice monoculture. We tested the hypothesis in two steps following Rahman (2003a, 2005) and Rahman and Thapa (1999). The first step was to create a farmer socio-environmental awareness index ($AvPI_i$) and afterwards apply Tobit and 'Propensity score matching' (PSM) approaches to evaluate the results. There are a few studies on farmer perceptions about sustainable technologies, a topic that has been reviewed by Tatlidil et al. (2009), and there do not appear to have been any studies that have compared the performance of rice–fish based IAA to rice monoculture systems in Bangladesh or elsewhere with similar considerations.

The remainder of this chapter is organized as follows. The next section discusses details the data and methodological approach used to analyse the data. In section three, farm level input use comparison and Meta review results and detail descriptive and econometric results are discussed under different subsection. Finally, section four derives policy implications and concludes the chapter.

5.2 Methods

5.2.1 Data

The study used cross-sectional data collected at the farm level as a follow-up survey effort conducted from July 2014 until January 2015 in 21 sub-districts from five districts in northern and northwestern Bangladesh (see Figure 1.2). To fulfil the objectives of this study a total of 494 farm households from the original panel households were selected. Among these sample households, 73 (14.78%) were rice–fish based IAA producers and all households had some experience in monoculture rice production as well as integrated rice–fish farming. These sample households were surveyed regarding socio-economic and environmental perceptions about both farming systems. The remaining 421 (85.22%) of the sample households practiced monoculture rice production only. Among the 421 rice monoculture farm households, 269 (64%) were indigenous and 152 (36%) were non-indigenous farm households. The 421 monoculture rice farmers were asked about the socio-economic and environmental perceptions about rice monoculture only. The survey effort also included detailed parcel level data on inputs and outputs of rice–fish based IAA and rice monoculture production.

5.2.2 Econometric framework

The decision of whether or not to adopt a new technology is based on two alternative formulations, one consists of the relative costs and benefits and the other is a random utility framework (e.g. Adesina and Zinnah, 1993; Adesina and Baidu-Forson, 1995; Baidu-Forson, 1999). Although these two formulations have slightly different motivations, they lead to the same set of econometric models (Barham et al., 2004). The factors affecting the utility or derivable benefits from a particular technology and that in turn affect the adoption of new technology can be explained by three main models or paradigms. These are: (i) the innovation-diffusion model; (ii) the economic constraints model; and (iii) the technology characteristics-user context model, also known as the adoption perception model (Adesina and Zinnah, 1993; Negatu and Parikh, 1999). Following Rahman (2003a, 2003b), we placed more emphasis on the adoption perception paradigm based on the hypothesis that, at the ex-post adoption stage an observable set of technology specific characteristics and farm specific socio-economic attributes will similarly influence farmer awareness (in this case farmer perceptions) of the positive or negative socio-economic and environmental impacts associated with the adopted technology (i.e. in this case rice monoculture and integrated rice–fish farming systems). Negatu and Parikh, (1999), and Rahman (2003a, 2003b, 2005) also examined ex-post adoption stage farmer perceptions.

Following the standard index model (e.g. Wooldridge, 2010b) we used a Tobit model to identify the determinants of farmers' socio-economic and environmental perceptions (awareness) of rice–fish based IAA and rice monoculture farming systems. The Tobit model can be expressed as follows (McDonald and Moffit, 1980):

$$y_i = X_i\beta + \mu_i; \text{ if } X_i\beta + \mu_i > 0$$
$$= 0 \text{ if } X_i\beta + \mu_i \leq 0 \qquad (1)$$
$$i = 0, 1, 2\ldots\ldots\ldots..n$$

where n is the number of observations, y_i is the dependent variable (mean farmer perception index = $AvPI_j$), Xi is a vector of independent variables, and μ_i is an independently distributed error term that is assumed to be normal with a mean of zero and a constant variance of σ^2.

The Tobit model formulated by Tobin (1958) has several advantages compared to alternative limited dependent variable models. Estimates are made using all observations, both those at the limit and those above it. For example, in this case farmers could be unaware of any socio- environmental impacts of rice–fish based IAA and rice monoculture systems even after practicing them. Potentially there are farmers with zero socio-environmental awareness at the limit. Thus the Tobit analysis is best suited for this type of censored data. The model permits the analysis of whether or not farmers have these perceptions and if so the extent of such perceptions. The Tobit model can also be used to further disaggregate coefficients to determine the effects of changes in the ith variable on change in the probability of perception and the expected change in intensity of perception. This can be expressed as:

$$E_y = F(z)E^*_y \qquad (2)$$

where E_y is the expected value of all observations, E^*_y is the expected conditional value above the limit, and $F(z)$ is the cumulative density normal distribution function at z where $z = X\beta/\sigma$. Following McDonald and Moffit (1980), differentiating Equation 2 with respect to the kth variable of X and additional transformation gives:[27]

$$(\delta E_y/\delta X_i)(X_i/E_y) = (\delta E^*_y/\delta X_i)(X_i/E^*_y) + \{\delta F(z)/\delta X_i\}\{X_i/F(z)\} \qquad (3)$$

In Equation 3 the total elasticity value can be disaggregated into two effects: (i) a change in the elasticity of intensity of an existing perception (change in perception) (i.e. for farmers above the limit); and (ii) change in the elasticity of perception (change in the probability of becoming aware) (i.e. the probability of being above the limit) (McDonald and Moffit, 1980; Adesina and Baidu-Forson, 1995; Rahman, 2003a).

In Equation 1 one of the X (explanatory variables) is the rice–fish based IAA adoption dummy, which is not random. Thus if there is an outcome (i.e. perception

---

27 Transformation was performed by multiplying by X_i/E_y and re-arranging Eq. 2.

index) difference between rice–fish based IAA adopters and non-adopters, it may be due to systemic differences rather than rice–fish based IAA adoption. This may cause a selection bias problem and the impact of rice–fish based IAA adoption on farmer perceptions would be overestimated or underestimated depending on the type of bias. There are different parametric and non-parametric econometric techniques[28] available to correct for potential bias in estimating such impacts (for details see Rosenbaum and Rubin, [1983] and Angrist, [2001]). To control for selection bias we used a popular non-experimental and non-parametric method (PSM)[29] as suggested by Rosenbaum and Rubin (1983, 1985). The PSM method creates the conditions of a randomized experiment in order to evaluate a causal effect as in a controlled experiment (Rosenbaum and Rubin, 1983). There are a number of steps in estimating the PSM approach. The first step of the PSM approach is to estimate farmer propensity scores (p score), which are estimated as follows:

$$P(X_i) = Pr(T_i = 1|X_i) \ (0<P(X_i)<1) \qquad (4)$$

where $P(X_i)$ is the propensity score estimated by a probit model that regresses T_i such as adoption status (rice–fish based IAA adopters = 1 and non-adopters = 0) on X_i, which is a vector of observed control variables.

After estimating farmer propensity scores the second step is to choose an appropriate matching algorithm for estimating the Average Treatment Effect on Treated (ATT). Following Becker and Ichino (2002) the ATT (in this case the impact of rice–fish based IAA adoption on farmer perceptions) is estimated as follows:

$$ATT = E\ [E\ \{Y_{1i}|\ T_i = 1, p\ (X_i)\} - E\ \{Y_{0i}|\ T_i = 0, p\ (X_i)\}|\ T_i = 1] \qquad (5)$$

where Y_{1i} and Y_{0i} are the two potential counterfactual outcomes of rice–fish based IAA adoption and non-adoption.

There are number of matching algorithms[30] that have been suggested in the literature to estimate relia ble ATT values. Here we have used several matching algorithms to pair IRFS adopters to similar non-adopters using the estimated propensity score in the first step. As such we estimated the ATT by using the PSM algorithm developed by Becker and Ichino (2002) with Stata 12.1.

---

28 See Mendola (2007) for advantages and dis-advantages of parametric and non-parametric approaches.
29 For details about the use and critics of PSM in econometric selection models see Heckman et al. (1997) and Smith and Todd (2005).
30 For comprehensive overviews on the various algorithms see Smith and Todd (2005) and Caliendo and Kopeinig (2008).

5.2.3 Dependent Variables: Farmer Socio-Environmental Awareness Index

By following the framework of Rahman and Thapa (1999) and Rahman (2003a, 2005) we constructed the farmer socio-environmental awareness index ($AvPI_j$), which is summarised in the Appendix in Table A.5.1 that includes details of the construction process. Data on farmer perceptions of the socio-environmental impacts of rice–fish based IAA and rice monoculture farming systems were collected in two steps. First, farmers were asked[31] about their opinions on a set of 24 specific socio-environmental impact indicators[32] associated with rice–fish based IAA and rice monoculture farming systems (I_j). If farmers recognized each of the impacts then a value of one was assigned and a value of zero was assigned if otherwise.

In the next step, farmers who recognized a particular impact were then asked to describe the extent (severity) of each impact indicator on a five-point Likert scale (1 to 5),[33] where one indicates least severe and five indicates a very high level of severity (R_m). A value of zero was assigned for indicators of impacts that were not recognized by a farmer. Then an overall perception index (AvPIi) score was calculated for each farmer by summing the severity values of each impact indicator (APIi) and dividing by the total number of impacts (24). The farmer perception index scores are reported in tables 5.3 and 5.4. The results show that perceptions of the adverse socio-environmental impacts of rice monoculture were more common among integrated rice–fish farmers than rice monoculture farmers (Table 5.4). Similarly, integrated rice–fish farmers reported that the adverse socio-environmental impacts of rice monoculture are comparatively higher than for integrated rice–fish systems (Table 5.3).

5.2.4 Independent Variables

The explanatory variables reported in Table 5.1 that were included in the Tobit model were expected to influence farmer perceptions regarding the socio-environmental impacts associated with rice–fish based IAA and rice monoculture

---

31 As integrated rice–fish farming system farmers had engaged with both rice–fish as well rice monoculture production they were asked to give their opinions on both farming systems, however rice monoculture farmers were only asked about rice monoculture systems because they lack direct experience with integrated rice–fish systems.
32 Selection of the list of indicators was based on the results of a literature review of similar studies in Bangladesh and elsewhere, and all of those indicators were validated before being incorporated into the questionnaire through pre-testing with farmers and experts.
33 This range captures the intensity of their opinions about each indicator.

systems. The inclusion of these explanatory variables was guided by relevant perceptions studies and theories. In addition, before the inclusion of the explanatory variables we checked for collinearity among them because it can lead to imprecise parameter estimates and difficulties in identifying the unique role of each explanatory variable on the dependent variable. Therefore we checked for the presence of multicollinearity using a correlation analysis and the variance inflation factor (VIF) before inclusion in the model.

The summary statistics in Table 5.1 describe the characteristics of the two farmer categories. The right-hand side of the Tobit model included a range of variables that potentially explain the determinants of farmer perceptions on the socio-environmental impacts of rice monoculture. On average rice–fish based IAA farmers were older and more educated than rice monoculture farmers. All of the rice–fish based IAA farmers were indigenous farmers, however, the rice monoculture farmers included both indigenous and non-indigenous farmers.

Table 5.1: Summary statistics of independent variables used in the regression

Variables	Type (definition and measurement)	Rice monoculture farmers		Integrated rice–fish farmers		Difference
		Mean	Std. Dev.	Mean	Std. Dev.	
Age of HH head	Continuous (age of the farmer in years)	49.17	13.09	51.93	13.54	−2.76*
Education of HH head	Continuous (Education level in years)	3.75	3.93	4.93	4.29	−1.18**
HH Ethnicity	Dummy (1 if indigenous, 0 if otherwise)	0.36	0.48	0.00	0.00	0.36***
Household size	Continuous (total household members)	4.68	1.59	4.88	1.74	−0.19
Main occupation of HH head	Dummy (1 if main occupation is agriculture, 0 if otherwise)	0.71	0.45	0.73	0.45	−0.01
Farm size	Continuous (total land area in decimals)	144.36	168.4	196.81	176.25	−52.44**
Off-farm income	Continuous (total annual off-farm income)	19504.55	26486.9	13234.96	20038.9	6269.59**
Extension services	Dummy (1 if farmer had access to extension services)	0.45	0.50	0.48	0.50	−0.03
Project participation	Dummy (1 if participated in AFP, 0 if otherwise)	0.38	0.49	1.00	0.00	−0.62***
Irrigation	Dummy (1 if farmer had access, 0 if otherwise)	0.72	0.45	0.79	0.41	−0.08
Infrastructure	Continuous (distance from home to sub-district in Km)	7.08	5.55	5.59	5.99	1.49**
Observations (N)		421		73		

Notes: * Significance at 10%; ** significance at 5%; *** significance at 1%.

Furthermore, the descriptive results presented in Table 5.1 show that there was a slight difference in household size between rice monoculture and rice–fish based IAA farmers that was not significant. Similarly, the main occupation of respondents from both groups did not differ significantly, with the majority of both groups (71–73%) reporting agriculture as their main occupation. On the other hand farm size between the two categories differed significantly. Rice–fish based IAA farmers had significantly larger farms than rice monoculture farmers. Conversely, rice monoculture farmers had significantly greater off-farm income than rice–fish based IAA farmers.

Table 5.1 also presents the results regarding extension services availability, which was nearly 50% of farmers for both groups. While 100% of the rice–fish based IAA farmers participated in the AFP,[34] only 38% of rice monoculture farmers had participated in the project. On the other hand, there was no significant difference between rice–fish based IAA and rice monoculture farmers regarding the access to irrigation. Table 5.1 also displays information on infrastructure variables for rice–fish based IAA and rice monoculture farmers. On average rice–fish based IAA farmers were located significantly closer to the nearest sub-district (*upazila*), which is typically the main market centre at the sub-district level.

5.3 Results and Discussion

5.3.1 Plot Level Comparison of Inputs Used for Rice–fish based Integrated Aquaculture-Agriculture and Rice Monoculture Farming Systems

Table 5.2 presents the actual plot level (not perception based) costs and benefits of rice–fish based IAA and rice monoculture production with a special focus on items with greater socio-environmental impacts. At the plot level the mean per hectare total labour cost varied by farming system, with rice–fish based IAA systems having greater labour needs, suggesting greater employment potential, especially for household labour. IAA systems also had greater employment potential for women. This is consistent with the findings of Nabi (2008) that rice–fish based IAA systems require more labour than rice monoculture, which were consistent with the findings of Halwart and Gupta (2004) and Sollow (2000). A more recent

---

[34] Participation in the AFP was used as a proxy for FFS because the project adopted the FFS approach to disseminate the prospective technologies and to train participants. See details about the AFP in Pant et al. (2014).

study in Bangladesh also reported similar findings (Dey et al., 2013). That study mentioned that some of the rice–fish based IAA activities such as feed preparation, fish feeding, and loss prevention are not strenuous and that women can more easily perform these functions than rice monoculture farming responsibilities. In addition, rice–fish based IAA systems create lean season employment opportunities for rural smallholders. Omidi-Najafabadi and Masjedi (2011) found that rice–fish based IAA systems provide off season employment to farm labourers. Prein (1998) found that rice–fish based IAA systems require a significant amount of labour, which is a challenge to the promotion and broader adoption of rice–fish based IAA in rural economies due to widespread labour scarcity.

Mean rice yields were significantly different across farming systems, with greater yields among rice monoculture systems. While the mean rice yields were higher for rice monoculture systems, the rice–fish based IAA systems had significantly greater total returns and gross margins. There have been many studies on rice yields of rice–fish based IAA systems relative to rice monoculture, most of which were experimental. Some found increased rice yields among rice–fish based IAA systems whereas others found no effects at all, and some found relatively lower rice yields. Thongpan et al. (1992) reported rice yields that were 1% to 29% lower among rice–fish based IAA systems. Rice yields in fields with fish were lower compared to fields without fish in China (Li et al., 1995) and the Philippines (Van Dam, 1990). Sollow (2000) also found that rice yields may be lower among rice–fish systems. Based on eight experiments in Vietnam, Vromant et al. (2002) concluded that fish production did not have any impact on rice yields. Xu and Guo (1992) also found that fish production had no significant effect on rice yields. Similar findings were reported by studies in Indonesia (Yunus et al. 1992; Hendarsih et al. 1994), Thailand (Taylor et al. 1988), the Philippines (Arce and Dela Cruz, 1979; Cagauan et al. 1994b), and Bangladesh (Yousuf et al., 1992). Controlled experiments in Vietnam found no significant differences in rice yields between plots with and without fish (Vromant et al. 1998; Rothuis et al. 1999). Experimental results from Bangladesh also indicated no significant differences in rice (grain and straw) yields between rice–fish and rice monoculture plots (Frei et al., 2007a). A survey in the Mekong River delta of Vietnam confirmed these experimental results (Rothuis et al. 1998b). Torres et al. (1992) found that the presence of fish did not have any impact on rice yields.

Based on rice yield data from 20 rice–fish systems in China, India, Indonesia, the Philippines and Thailand, Lightfoot et al. (1992) found that mean rice yields were greater among rice–fish based IAA systems relative to rice monoculture. Relatively higher rice yields have been reported for rice–fish based IAA systems in multiple studies (Hofstede and Ardiwinata, 1950; Sinhababu et al., 1983; Cagauan et al,. 1994a; Wu, 1995; Cagauan, 1999; Tsuruta et al., 2011). Based on a review of rice–fish based IAA systems in China, Kangmin (1988) indicated similar findings. Some early studies also reported that integration of fish production into rice cultivation areas increased rice yields (Satari, 1962; Li, 1986). Increased rice yields were also reported for integrated rice–fish production in Guyana and Suriname (Geer et al., 2007). Mukhopadhyay et al. (1992) reported 1–11% increases in rice yields in India, Syamsiah et al. (1992) reported 6.6% yield increases in Indonesia, and Yousuf et al. (1992) reported up to 14% rice yield increases in Bangladesh under integrated production efforts. Mishra and Mohanty (2004) and Mohanty et al. (2004) also reported that the rice–fish integration increased rice yields by 8–15%. Ahmed and Garnett (2011) reported postive effects of rice–fish integration on rice yield in Bangladesh.

Table 5.2: *Comparative farm level socio-environmental benefits and costs of rice–fish based integrated aquaculture-agriculture and rice monoculture production*

Comparative items	Unit*	Rice–fish	Rice monoculture	Differences
Total variable cost	BDT	133414.40	87477.47	-45936.98***
Total labour	Labourer Days	287.57	209.46	-78.11***
Total male labour	Labourer Days	237.92	181.90	-56.02***
Total female labour	Labourer Days	49.65	27.57	-22.09***
Chemical fertiliser use	Kilogram (kg)	359.37	495.37	136.00**
Organic fertiliser use+	kg	102951.60	44116.87	-58834.78***
Concentrated pesticide use	kg	0.70	3.70	3.09***
Liquid pesticide use	Millilitre (ml)	300.76	672.66	371.89**
Pesticide application rate	Number of applications	1.15	1.96	0.81***
Proportion of farmers using pesticides	%	64.58	82.64	–
Cost of pesticides	Bangladeshi Taka (BDT)	1041.72	2101.77	1060.05 ***
Labour for pesticide application	Labourer Days	2.15	5.70	3.56***
Irrigation costs	BDT	11639.09	10228.21	-1410.88
Irrigation rate/frequency	Number of times	11.50	7.55	-3.95***
Total number of labour days required for weeding	Labourer Days	28.89	40.86	11.97 **
Weeding frequency	Number of times	1.25	1.40	0.15
Cost of weeding	BDT	5974.43	7815.11	1840.68*
Total return	BDT	258595.90	135985.00	-122610.9***
Rice output	Mound†	159.82	233.19	73.38***
Gross margins	BDT	172247.95	75279.66	-96968.29***

Notes: *All values are expressed as per hectare and per year; † one maund equals 40 kg; + organic fertiliser for rice–fish systems includes organic feed also; * significance at 10%; ** significance at 5%; *** significance at 1%.

Some studies indicated that rice production may increase or decrease depending on the fish species and management practices involved. The effects of different fish species and management practices on rice yields requires more research. Almost all of the studies mentioned above (experimental and survey based) showed comparatively higher returns for rice–fish based IAA systems than rice monoculture when total returns, gross margins, or net returns were evaluated (Arevalo, 1987; Dewan, 1992; Das et al., 2002; Hossain et al., 2005; Ofori et al., 2005; Frei et al., 2007a; Ahmed and Garnett, 2011; Dey et al., 2013). The gross margin results in this study revealed that rice–fish based IAA systems had comparatively higher gross margins than rice monoculture systems (Table 5.2).

The information on the environmental impacts related to inputs like chemical fertiliser use (Table 5.2) shows that rice monoculture system farmers applied significantly more fertiliser than rice–fish based IAA system farmers. Organic fertiliser use, especially the use of cow dung in rice–fish systems was significantly higher than for rice monoculture systems. Similar to chemical fertilisers use, pesticide application quantities and costs, including concentrate and liquid application in rice monoculture, were significantly higher than for integrated rice–fish systems. More than 82% of rice monoculture farmers reported that they used pesticides, while this figure was only about 65% for rice–fish systems. Similarly the frequency of pesticide applications and the labour requirements for pesticide applications by rice monoculture farmers were significantly higher than for rice–fish farmers. Irrigation costs did not differ much between the two farming systems, but irrigation frequency by rice–fish farmers was significantly higher than for rice monoculture. Halwart and Gupta (2004) reported that rice–fish systems require more water than rice monoculture systems. Weeding costs reported by rice monoculture farmers were higher than for rice–fish systems. Weeding frequency and the number of labour days dedicated to weeding among rice monoculture farmers were also higher than for rice–fish systems. Rice monoculture systems required significantly more environmentally hazardous inputs than rice–fish systems. Rice–fish systems produced higher socio-economic benefits than rice monoculture systems. Except for some variation, rice–fish farmer perceptions (section 5.3.2) were consistent with the plot level evidence for these two farming systems.

According to Wilson and Tisdell (2001) the current levels of pesticide application for rice production are believed to be either economically or environmentally unsustainable. Similarly, Berg (2002) found that increased reliance on pesticides in rice production is unsustainable because it contributes to outbreaks of insect pests, the development of pesticide resistant pests, and causes many human and environmental health problems. Some studies have claimed that although agricultural intensification in South Asia has significantly increased food production, it has also negatively affected the physical environment through the degradation and depletion of natural resources and the loss of genetic and biological diversity (Alauddin and Tisdell, 1998; Alauddin and Quiggin, 2008). Based on soil test results from different locations between 1967 and 1995, Ali et al. (1997) reported that soil fertility levels in Bangladesh are declining. A recent study in Bangladesh by Dey and Haq (2009) reported that in the case of boro rice growers, 55% of farmers applied urea above the recommended levels. Overapplication of chemical fertilisers causes environmental damage by negatively affecting soil and water quality. In addition they also indicated that pesticide related public health, water and environmental problems are very common in Bangladesh. Klemick and Lichtenberg (2008) found major problems associated with the effects of the GR, particularly environmental and human health problems caused by pesticides. Pesticides can both aggravate pest problems and disrupt arthropod food webs (Way and Heong, 1994; Schoenly et al., 1996), and may reduce long term farmer productivity because of related health problems (Rola and Pingali, 1993; Antle and Pingali, 1994; Huang et al., 2005).

Findings from Arunachal Pradesh, India indicate that rice–fish integration reduces the use of fertilisers, pesticides and herbicides at the field level (Saikia and Das, 2008). Based on experiences of a project in Guyana and Suriname, Geer et al. (2007) also reported similar findings and that reductions were even greater when rice–fish adoption was accompanied by a FFS approach. Other studies have also indicated that rice–fish integration reduces fertiliser, pesticide, and herbicide use (Berg, 2002; Noorhosseini and Radjabi, 2010; Noorhosseini-Niyaki and Allahyari, 2012). Some studies have claimed that integrated rice–fish farming can be an important element of IPM for rice crops (Berg, 2001; Halwart and Gupta 2004; Hilbrands and Yzerman, 2004; Frei et al., 2007b). Based on a field survey and experiment, Xie et al. (2011a) found that rice–fish systems require 68% less pesticides and 24% less chemical fertiliser than rice monoculture systems. Lu and Li (2006) reviewed rice–fish farming systems in China and found that pesticide applications can be lowered by 50% through the integration of fish production into rice fields.

Halwart et al. (1996) found that fish are potentially important predators of some rice pests. Other studies have also reported that because many fish species feed on aquatic organisms (e.g. insects, snails, weeds) they act as biological control agents (Kangmin, 1988; Lightfoot et al., 1992; Fernando, 1993; Dela Cruz, 1994; Cagauan, 1995; Halwart, 1995, 1998; Ichinose et al., 2002; Frei et al., 2007b). Based on field trials in India, Patra and Sinhababu (1995) reported reductions of weed biomass by 39% in rice–fish systems. Similarly Rothuis et al. (1999) reported a 100% reduction of submerged and floating weeds and Frei and Becker (2005a) noted a complete elimination of filamentous algae in rice–fish systems.

Fish culture in rice fields improves soil quality and fertility, as well as dissolved oxygen concentrations (Kangmin, 1988; Nie et al., 1992). Due to more intensive management of water flow and the fertilising effect of fish feces, nutrient availability (e.g. nitrogen, phosphorus, calcium, and magnesium concentrations) and the mineralization of soil nutrients were increased, and the volatilization of nitrogen was reduced in rice–fish systems (Satari, 1962; Li, 1986; Lightfoot et al., 1992; Langhu, 1995; Vromant et al., 2002). Based on their review of rice–fish farming systems in China, Lu and Li (2006) found that due to increased nitrogen fixation in rice–fish systems, organic matter, total nitrogen and total phosphorus in the soil increased by 15.6–38.5%. Other studies have also indicated that rice–fish systems improve soil fertility by increasing the availability of nitrogen and phosphorus in the soil (Giap et al., 2005; Dugan et al., 2006).

Lu and Li (2006) found that rice–fish systems reduce CH_4 emissions by nearly 30% compared to rice monoculture farming. Other authors have also indicated fish feeding actitvites add oxygen to the soil, which in turn is expected to mitigate methane production (Lightfoot et al., 1992; Langhu, 1995; Vromant et al., 2002). In contrast Frei and Becker (2005b) and Frei et al. (2007c) reported that the presence of fish boosts methane emissions from rice–fish systems compared to rice monoculture systems. Datta et al. (2009) also found that based on the total greenhouse gas emissions (in terms of CO_2 equivalents) that global warming potential is significantly higher for integrated rice–fish plots relative to rice monoculture plots.

Fernando and Halwart (2000) reported that in rice–fish systems fish may control the propagation of malaria vectors. Similar effects were reported by Jianguo and Dashu (1995) and Neng et al. (1995) in China, and by Lee and Lee (2003) in Korea. Matteson (2000) also indicated that fish can help control malaria and water-borne diseases. Xie et al. (2011b) showed that some traditional rice varieties can be preserved through rice–fish systems because those varieties were traditionally cultivated using integrated rice–fish systems in China. Lu and Li (2006) found that rice–fish systems in China contributed to the biodiversity of agricultural landscapes and the conservation of both rice varieties and fish species. Koohanfkan and Furtado (2004) found similar effects of rice–fish systems.

5.3.2 Comparison of Farmer Perceptions

5.3.2.1 Rice-fish based IAA farmer perceptions of the socio-environmental impacts of rice monoculture and integrated rice-fish systems

First we tested the hypothesis that, based on farmer perceptions, integrated rice–fish farming systems offer superior alternatives to rice monoculture in terms of socio-environmental criteria. Table 5.3 presents a summary of the integrated rice–fish farmer perceptions on the socio-environmental impacts of rice monoculture and integrated rice–fish farming systems. On average only 9% of the rice–fish based IAA system farmers identified different adverse socio-environmental impacts of integrated rice–fish farming systems, while more than 54% identified different adverse impacts of rice monoculture. Similarly, the mean impact severity values indicate similar results (i.e. the adverse socio-environmental impacts of rice monoculture are perceived as more severe than those of rice–fish based IAA systems). Thus based on farmer perceptions, integrated rice–fish farming systems performance is superior than rice monoculture in terms of socio-environmental impacts.

At the individual level the socio-environmental impact indicators of integrated rice–fish systems most reported by farmers were the increased need for inputs (19.18%), followed by increased use of child labour and contamination of water sources, and the remaining impact indicators were reported by less than 10% of the farmers. The most commonly reported socio-environmental impact indicators of rice monoculture systems were increased pest problems and increased need for inputs (both were 82.19%), followed by decreasing production over time, reduced fish harvests, greater soil compaction, increased crop or fish disease problems, reductions in the quantity of beneficial organisms, reduced soil fertility, and the remainder of the indicators the response percentages were near 50% or below.

Based on reported severity levels almost all impact indicators were percieved to be higher for rice monoculture than for rice–fish based IAA systems. The most severe perceived impact of rice monoculture was reductions in the quantity of beneficial organisms, followed by decreasing production over time, increased insect/pest problems, increased need for inputs, reduced soil fertility, greater soil compaction, depletion of water tables, and increased crop or fish disease problems. Although farmers indicated their awareness of the generally more adverse socio-environmental impacts of rice monoculture relative to rice–fish based IAA systems, their perceptions were limited to directly observable impacts on their farm fields. Perceptions about indirect impacts such as reduced organic matter in soils, increased soil acidification, and increased soil salinity were less represented as both impacts and severity values.

Table 5.3: Rice–fish based IAA farmer perceptions of the impacts of rice monoculture and integrated rice–fish systems in Bangladesh

Socio-environmental impacts	Percentage of integrated rice–fish farmers reporting impacts		Mean perceived severity scores of perceived impacts*	
	RF-IAA (%)	RM (%)	RF-IAA	RM
Adversely affects human and animal health	5.48	42.47	2	3.26
Reduces soil fertility	5.48	69.86	1.75	3.59
Reduces fish harvests	2.74	78.08	2	3.35
Increases disease of crops or fish	1.37	75.34	1	3.56
Increases soil compaction	5.48	76.71	1.25	3.59
Increases insect/pest problems	9.59	82.19	1.29	3.78
Reduces soil organic matter	6.85	60.27	2.8	1
Increases soil acidification	4.11	38.36	2.33	1
Increases soil erosion	10.96	52.05	3.5	3.08
Increases soil salinity	6.85	26.03	2	2.42
Reduces biodiversity	10.96	54.79	2.25	3.05
Reduces quantity of beneficial organisms	13.7	73.97	2.2	4.02
Contaminates water	16.44	50.68	2.38	3.54
Depletes ground water	10.96	50.68	1.88	3.57
Creates water logging	13.7	31.51	2.1	3.04
Increases need for inputs	19.18	82.19	3.36	3.68
Reduces productivity over time	10.96	81.54	2.38	3.89
Inefficient fertilisation	6.85	26.03	2.2	3.16
Reduces labour requirements	12.33	19.18	2.11	3.29
Reduces role of women	10.96	26.03	2	3.37
Increases child labour	17.81	52.05	3.23	3.08
Reduces dietary diversity	2.74	58.9	2	3.05
Contributes to local irrigation water conflict	8.22	49.32	2.83	3.33
Contributes to local land conflict	6.85	42.47	3	3.03
Mean	9.19	54.20	2.24	3.16

Notes: RF-IAA = rice–fish based IAA system; RM = rice monoculture; * 1 = least severe; 5 = most severe.

5.3.2.2 Comparison of perceptions of the socio-environmental impacts of rice monoculture among integrated rice–fish farmers and rice monoculture farmers

Table 5.4 presents the results of perceived rice monoculture socio-environmental impacts among both rice monoculture and rice–fish based IAA farmers. On average, higher percentages of rice–fish based IAA farmers reported adverse socio-environmental impacts of rice monoculture than rice monoculture farmers. Similarly, the mean severity values of impacts reported by rice–fish based IAA farmers (3.16) were higher than those of rice monoculture farmers (3.00).

Seventeen of the 24 individual impact indicators were recognised for rice monoculture systems by comparatively higher percentages of rice–fish based IAA system farmers than rice monoculture farmers. The adverse impacts of rice monoculture that were recognised by the highest percentage of rice monoculture famers are reduced soil fertility, followed by increased insect/pest problems, increased crop or fish diseases, reduced soil organic matter, production declines over time, and decreased quantity of beneficial organisms. Similarly, the adverse impacts perceived by the highest percentages of rice–fish based IAA system farmers were increased insect/pest problems and increased input requirements, followed by production declines over time, reduced fish harvests, compacted/hardened soils, increased crop or fish diseases, reduced quantity of beneficial organisms, and reduced soil fertility. In terms of severity, both types of farmers had similar perceptions of the greater severity of rice monoculture impacts relative to IRFFS systems, but more of the rice–fish based IAA farmers perceived more severe impacts than rice monoculture farmers. Severity scores also reveal that farmers perceived the severity of rice monoculture impacts that were directly observable more than unobservable impact indicators.

Table 5.4: Rice–fish based IAA system and rice monoculture farmer perceptions on the socio-environmental impacts of rice monoculture systems in Bangladesh

Socio-environmental impacts	Percentage of farmers reporting impact		Mean severity scores of perceived impacts	
	RM farmers (%)	RF-IAA farmers (%)	RF-IAA farmers	RM farmers
Adversely affects human and animal health	52.73	42.47	3.26	3.64
Reduces soil fertility	83.61	69.86	3.59	3.32
Reduces fish harvests	52.26	78.08	3.35	3.02
Increases crop or fish disease	71.50	75.34	3.56	3.19
Causes soil compaction	60.10	76.71	3.59	3.12
Increases insect/pest problems	81.47	82.19	3.78	3.36
Reduces soil organic matter	71.5	60.27	1.00	1.00
Increases soil acidification	33.02	38.36	1.00	1.00
Increases soil erosion	29.45	52.05	3.08	2.65
Increases soil salinity	18.29	26.03	2.42	2.51
Reduces biodiversity	46.79	54.79	3.05	3.39
Reduces quantity of beneficial organisms	67.46	73.97	4.02	3.68
Contaminates water	52.97	50.68	3.54	3.21
Depletes water table	37.77	50.68	3.57	3.42
Creates water logging	27.32	31.51	3.04	2.95
Increases need for inputs	61.28	82.19	3.68	3.25
Reduces productivity over time	67.93	81.54	3.89	3.51
Inefficient fertilisation	38.48	26.03	3.16	3.28
Reduces labour requirements	37.29	19.18	3.29	3.29
Reduces role of women	38.48	26.03	3.37	2.92
Increases child labour	38.72	52.05	3.08	2.39
Reduces dietary diversity	42.99	58.9	3.05	2.98
Contributes to local irrigation water conflict	40.86	49.32	3.33	3.42
Contributes to local land conflict	32.78	42.47	3.03	3.41
Mean	49.38	54.20	3.16	3.00

Notes: RF-IAA = rice–fish based IAA system; RM = rice monoculture; * 1 = least severe; 5 = most severe.

5.3.3 Determinants of Farmer Awareness of the Socio-environmental Impacts of Rice Monoculture: Tobit and PSM Analyses

Table 5.5 presents the Tobit model results[35] of the determinants of farmer perceptions about rice monoculture impacts based on the socio-environmental perception index (AvPIi). The AvPIi is a composite index that considers 24 impact indicators. The elasticity values computed using Equation 3 are also reported in Table 5.5. Variables that significantly affected the perception and the intensity of perception are: farmer ethnicity, off-farm income, AFP participation, rice–fish based IAA adoption, distance to the nearest sub-district (as an infrastructure access proxy variable), and its squared term. The signs for the relationships between these variables and the outcomes are consistent with expectations except for the infrastructure proxy variable, which was positive, however, the relationship with its squared term was negative as expected. Farmers living nearer to city centres were more likely to perceive socio-environmental impacts because they are more exposed to information and opportunities to exchange information. The significant positive relationship between distance and awareness, and the significant negative relationship between awareness and the squared distance term indicate that beyond a certain distance, greater distance limits farmer perceptions. Negatu and Parikh (1999) and Rahman (2003a) found that distance to town (as a proxy for infrastructure access) was an important explanatory variable affecting perception.

The elasticity estimate results show that marginal changes in the explanatory variable increase the probability of perceiving socio-environmental impacts slightly more than changes in the intensity of perceived impacts. The elasticity estimates, however, show highly inelastic responses to changes in the explanatory variables. Project participation and infrastructure access had the highest impacts on the probability of perception and the intensity of perception. The total elasticity value for project participation is 0.10, which was divided between an elasticity of perception probability (0.052) and an elasticity of intensity to perception (0.048). A 10% increase in project activity participation is expected to result in an increase of about 1% in the perception of and the intensity of perception of the socio-environmental impacts of rice monoculture. The probability of perceiving impacts will increase by more than 0.5%, while the intensity of perception will increase by 0.48%. Variables such as off-farm income, farmer ethnicity, and rice–fish based IAA adoption were all expected to significantly increase the perception of and the intensity of perception as expected. Studies on the environmental impacts of modern agricultural technology diffusion in Bangladesh have reached similar results (Rahman, 2003a).

---

35 Stata version 12.1 was used for the analyses.

Table 5.5: Tobit model results for determinants of farmer perceptions of the socio-environmental impacts of rice monoculture in Bangladesh

Farmer socio-environmental perception index (SEPI) variables	Coefficients (Standard Error)	Elasticity of	
		Perceiving impacts (probability)	Intensity of perception (change in perception)
Farmer age	0.003 (0.004)	0.053	0.049
Farmer education level	0.008 (0.013)	0.013	0.012
Farmer ethnicity	0.264* (0.144)	0.032	0.030
Family size	−0.025 (0.032)	−0.045	−0.042
Farmer's main occupation	−0.116 (0.117)	−0.032	−0.030
Farm size	0.000 (0.000)	0.007	0.007
Off-farm income	0.000** (0.000)	0.037	0.035
Access to extension services	0.029 (0.102)	0.005	0.005
AFP participation	0.280** (0.138)	0.052	0.048
Integrated rice–fish farmer†	0.305* (0.160)	0.018	0.016
Irrigation use	0.146 (0.115)	0.042	0.039
Access to infrastructure	0.050** (0.023)	0.049	0.046
Infrastructure access squared	−0.002** (0.001)	–	–
Constant	1.831*** (0.297)	–	–
Log likelihood	−749.19		
LR chi^2(13)	23.44		
Prob > chi^2	0.0367		
Pseudo R^2	0.015		
Number of observations	494		

Notes: † Dummy variable (1 if respondent is an rice–fish based IAA farmer); *** significance at 1%; ** significance at 5%; * significance at 10%

The dummy explanatory variable 'rice–fish based IAA farmer or not' was positive and significant, indicating that the adoption of rice–fish based IAA significantly increases the probability that a farmer perceives the socio-environmental impacts of rice monoculture (Table 5.5). However, adoption of rice–fish based IAA is likely subject to an endogenous selection bias problem. Thus to check the robustness of rice–fish based IAA adoption as a significant determinant of farmer perceptions about the socio-environmental impacts of rice monoculture we used the PSM approach already discussed in the econometrics section. The ATT results[36] using a different algorithm in the PSM approach are presented in Table 5.6. The positive

---

36 Before ATT estimation it was necessary to estimate propensity scores, but the results are not reported here to save space.

and significant ATT result indicates that rice–fish based IAA adoption is positively related to farmer perceptions about the adverse socio-environmental impacts of rice monoculture. More specifically, farmer perceptions about the adverse socio-environmental impacts of rice monoculture would be 40–70 points (depending on the matching algorithm) higher if farmers adopt rice–fish based IAA relative to non-adopters (i.e. rice monoculture farmers).

Table 5.6: ATT results from the alternative matching algorithm

Outcome	Matching Algorithm				
	Kernel (BW=0.03)	Kernel (BW=0.06)	Nearest-neighbour (NN=1)	Nearest-neighbour (NN=5)	Stratification method
Farmer perception index (AvPIi) of adverse socio-environmental impacts of rice monoculture	0.65**	0.56***	0.66**	0.43**	0.70**

Notes: ATT estimates of the nearest neighbour matching were obtained by applying the 'nnmatch' command using the bias adjustment option in Stata 12.1 (Abadie et al., 2004) and ATT estimates of kernel matching (based on the Epanechnikov kernel) and stratification methods were obtained by implementing the 'attk' and 'atts' commands (Becker and Ichino, 2002); bootstrapped standard errors (number of replications = 100) were also estimated for kernel and stratification matching estimators; in the kernel and stratification matching estimators the common support condition and balancing condition are imposed and satisfied and the matched sample includes 73 rice–fish adopters and 152 non-adopters (rice monoculture farmers); a balancing test[37] was used to check the quality of the matching based on the 'pstest both' command and the balancing condition was sufficiently satisfied; the distributions of the estimated propensity scores were checked using the 'psgraph' command and to ensure substantial overlap in their propensity score distributions (Leuven and Sianesi, 2003); *** significance at 1%; ** significance at 5%; and * significance at 10%

5.4 Conclusions and Policy Implications

Over the last few decades agricultural production has increased immensely in many parts of the world through the increased use of GR technologies and practices. The overuse and/or inappropriate use of agrochemicals and mono-culture crop production, especially intensified rice production, have led to the contamination and depletion of water resources, the loss of genetic and biological diversity, soil degradation, and human health problems. Thus intensive rice monoculture production cannot provide a sustainable food production solution due to long-term environmental impacts. In this situation, finding alternative

---

37 The balancing test and psgraph results are not reported here.

methods to produce more food and fiber on a sustainable basis for growing populations has become a serious challenge for developing countries like Bangladesh. In this sustainable intensification paradigm[38] 'climate smart' agricultural practices[39] are gaining popularity in the developing world.

This study investigated rice–fish based IAA as a potentially sustainable to rice monoculture. Specifically we investigated one of the least explored issues associated with the diffusion of farming system innovation, specifically the comparative socio-environmental impacts of rice–fish based IAA and rice monoculture by comparing farm level inputs and outputs, an extensive literature review, and an analysis of farmer perceptions and their determinants. Literature on the relative socio-environmental performance of rice–fish based IAA and rice monoculture based on farmer perceptions and their determinants is unavailable.

We investigated this empirically using two econometric models that correct for selection bias that influences rice–fish based IAA adoption and consequently farmer perceptions. Firstly, we compared farm level input use between rice–fish based IAA and rice monoculture (Table 5.2). The findings indicate that rice–fish based IAA farmers use less environmentally hazardous inputs and more environmentally benign inputs compared to rice monoculture farmers. The results also indicate that rice–fish based IAA is a socio-economically rational production system. The findings of the literature review also support and validate the farm level results. Secondly, we investigated farmer perceptions of the socio-environmental impacts of rice–fish based IAA and rice monoculture. Rice–fish based IAA farmers were asked about both types of farming systems because they had practical experience with both systems, but rice monoculture farmers were only asked about rice monoculture impacts because they lacked experience with rice–fish based IAA. Both types of farmers were well aware of the socio-environmental impacts of respective farming systems that they are exposed to. Rice–fish based IAA farmers perceived that rice–fish based IAA has relatively less negative socio-environmental impacts relative to rice monoculture (Table 5.3). The adverse socio-environmental impacts of rice monoculture were perceived by both types of farmers, but rice–fish based IAA farmers were more aware of the

---

38 According to the Montpellier Panel Report (2013) sustainable intensification "is about producing more outputs with more efficient and prudent use of all inputs—on a durable basis—while reducing environmental damage and building resilience, natural capital, and the flow of environmental services."
39 According to FAO (2010b) climate-smart agriculture (CSA) "sustainably increases productivity, resilience (adaptation), reduces/removes GHG (mitigation), and enhances the achievement of national food security and development goals."

adverse socio-environmental impacts of rice monoculture relative to rice monoculture farmers (Table 5.4). All results are based on a perception index created using farmer responses on different socio-environmental indicators and their perceived intensity. Most farmer perceptions remained limited to directly observable impacts such as changes in the occurrence of diseases and pests, relative soil differences, and differences in the quantity or kinds of beneficial organisms, and fish harvests, etc. Thus farmer perceptions of socio-environmental impacts and secondary evidence confirm the farm level comparison results (Table 5.2).

Thirdly, we examined the factors affecting the perception of socio-environmental impacts of rice monoculture by both type of farmers using Tobit and PSM approaches. The Tobit model results show that farmer ethnicity (indigenous or non-indigenous), off-farm income, AFP participation, rice–fish based IAA adoption, and distance to nearest sub-district (as a proxy for infrastructure access) are the major determinants of farmer perceptions of the adverse socio-environmental impacts of rice monoculture (Table 5.5). The PSM results further confirm rice–fish based IAA adoption as a major determinant of farmer perceptions (Table 5.6). Thus promotion of rice–fish based IAA, developing better infrastructure, and using FFS approaches to training[40] will likely help farmers develop their perceptiveness of the socio-environmental impacts of rice monoculture, which in turn may reduce the use of environmental hazardous inputs and increase interest in more environmentally benign and sustainable production technology.

The variables or indicators used in these analyses are not hard indicators (e.g. direct measurements of the physical or chemical properties of soils, water, or farmer health or productivity, etc.,). We have tried, however, to address these issues by using different indicators based on perceptions and a plot level comparison of input use using robust estimation techniques. Future research efforts could apply a conceptual framework that more fully accounts for those considerations (i.e. hard environmental indicator data), which would be useful for validating the findings of this study. Although our findings indicate that rice–fish based IAA adoption is a robust determinant of farmer awareness of the socio-environmental impacts of rice monoculture, we could not determine if this heightened perception leads to more judicious use of environmentally hazardous inputs or greater use of environmentally benign inputs. Rice–fish based IAA and rice monoculture

---

40 The 'project participation' dummy variable in the Tobit model had a significant and positive relationship with farmer awareness of the socio-environmental impacts of rice monoculture systems. Project participants received training on farming systems through the AFP using the FFS approach (for details about the AFP see Pant et al., 2014).

plot level input information from the same farmers will be essential to such an analysis. It is hoped that the results of this study can contribute to the development of a more comprehensive and sustainable agricultural development strategy in the developing world.

Chapter 6: Summary, Conclusion— Policy Implications and Further Research Needs

The discussion in this chapter reflects upon and highlights some of the most important findings that emerged from the analyses presented in the preceding chapters. The first section summarises the key findings from each chapter. The second section includes an overall conclusion of the study and draws policy implications based on some of the most fundamental research findings. The final section identifies a research and policy agenda for sustainable agricultural development in Bangladesh and elsewhere through intensification options like IAA.

6.1 Research Summary

Agricultural sectors provide income, livelihood options, and food security to the majority of the developing world, including Bangladesh, but currently agriculture faces great challenges to the production of sufficient food in a sustainable manner due to an increasing global population, limited land and water resources, and the effects of global climate change (Xie et al., 2011a). Sustainable agricultural intensification options are needed to meet those challenges.

Bangladesh is one of poorest and most densely populated countries in the world. Agriculture is still the backbone of the country's economy and is the major source of rural livelihoods, food security, and dietary nutrition. To meet the growing demand for food with limited land and water resources, Bangladeshi agriculture has adopted GR intensification strategies and made substantial progress. Such intensification strategies are highly criticized, however, due to sustainability concerns associated with negative environmental and human health impacts (Gupta et al., 1996; Klemick and Lichtenberg, 2008). There is an urgent need for more sustainable intensification options, especially for the production of staple foods. One such potential option is IAA, a traditional agricultural system that can be matched to local conditions and farming knowledge, and that integrates various farm enterprises by using diverse on-farm resources and organic inputs. Since the 1980s WF has been working on participatory development and promotional efforts for IAA in Bangladesh and many other Asian and African countries (Dey et al., 2010; Jahan and Pemsl, 2011; Xie et al., 2011a).

This socio-economic study on the determinants of participation in and impacts of IAA value chain development in Bangladesh had the following specific objectives:

o To evaluate the overall performance of IAA value chains in Bangladesh
o To identify the key determinants of IAA value chain participation dynamics.
o To assess the welfare impacts of IAA value chain participation dynamics.
o To evaluate the socio-environmental performance of rice–fish based IAA relative to rice-monoculture systems.

The analyses were based on panel data from a three-wave survey conducted in 2007, 2009, and 2012 by the WF Bangladesh programme and the author using structured questionnaires to interview members of sample households. The 2012 survey effort included additional questions to elicit information specific to the analyses presented in chapters two and five. Data were collected from IAA value chain participators and non-participators in the northern and northwestern regions of Bangladesh (see sample details in Table 1.1). IAA value chain participants include those in production activities such as rice–fish and pond fish based IAA farmers, cage culture, community-based pond aquaculture producers, and fingerling producers, as well as supporting value chain actors such as fish and fingerling traders, and fishermen.

Financial performance of the value chain actors was measured using a gross margin analysis. We further investigated whether the integration of fish into rice production would improve profitability sufficiently to justify a farming system improvement programme. Partial budget analyses were conducted for two rice production systems; a conventional system based on GR strategies (monoculture model) and diversification through aquaculture (integrated farm model). We also explored the internal and external factors affecting integrated rice–fish systems that affect the adoption and diffusion of IAA. We applied a SWOT analysis to better understand relevant issues including policy and institutional level strengths, weakness, opportunities and threats that can affect future strategy building regarding rice–fish technology adoption and diffusion. The findings indicate that integrated rice–fish systems offer considerable potential for increasing overall agricultural productivity and farm incomes in Bangladesh. The results also show that there are opportunities for extremely poor and landless households to participate in the rice–fish based IAA value chains in a profitable manner. There do not appear to have been similar efforts using detailed cost and return of rice–fish based IAA value-chains in Bangladesh or other countries with considerable potential for integrated rice–fish production. The findings demonstrate that rice–fish based IAA can create profitable business opportunities along the value-chain and employment opportunities for women, especially in production related activities. The partial budgeting analysis results corroborated that integrated rice–fish systems offer an economically competitive alternative to rice-monoculture in Bangladesh.

As a technologically innovative agricultural approach, integrated rice–fish production faces a number of significant challenges and has exhibited limited growth in Bangladesh. Experimentation by some innovative farmers, private efforts by WF, and NGOs like CARE are key drivers of efforts to mainstream integrated rice–fish production into agricultural development in Bangladesh. There is a virtual lack of governmental support to rice–fish farmers and overall value chain development in the country. The high start-up costs of integrated rice–fish production related activities in terms of land, labour, fingerlings, feed, and the required modifications to existing rice production areas are major constraints to increased and 'pro-poor' adoption and diffusion. In the short run, non-production rice–fish value chain actors have fewer entry barriers, and if combined with rice–fish farming, the benefits can be significantly greater over the long run for poor farmers despite high initial costs. The traditional strength of the IAA along with abundant water, fertile soils, strong research and extension institutions, expanding infrastructure, and favourable government policies for increasing private-sector participation outweigh by its weakness and related threats. Value-chain analysis has not been widely used to assess the performance of integrated farming systems in general, and for integrated rice–fish production in particular, for the purpose of promoting its further development.

The quantitative analyses presented in Chapter 3 investigated the IAA value chain participation dynamics of actors with respect to IAA production and non-production (up and down stream) activities using a range of cross-sectional and panel econometric approaches. The factors that influence the both IAA value chain participation and subsequent abandonment have rarely been addressed in technology adoption literature. Initially we used the panel data to expand the typical comparison of 'participators'/'adopters' and 'non-participators'/'non-adopters' to a more nuanced multinomial logit analysis of 'non-participators,' 'participators,' and 'dis-participators.' This revealed the differences among participators categories (both continuous participators and dis-participators) and non-participators as well as differences between continuous 'participators' and 'dis-participators.' The results revealed significant differences among participator categories (continuous participators and dis-participators) and non-participators, but few differences between 'participators' and 'dis-participators,' especially between production process participators and dis-participators, where dis-participation rate was comparatively lower relative to up and downstream IAA value chain actors. We then used the panel data to estimate three RE logit models that control for fixed and random omitted variables and endogenous regressors. The results found significant effects of omitted variables and confirm the robustness of the literature consensus that higher education, larger families, access to extension services, CBO membership, and access to market information are associated with participation in IAA value chain activities (or technology adoption in general). These results were also shown to be robust with respect to potential endogeneity of the four household characteristics.

Overall, the empirical analyses of IAA participation dynamics have led to the following conclusions about the future adoption and diffusion of IAA technologies in Bangladesh and socio-economically, environmentally, and institutionally similar countries. First, non-participants are very distinct from members of the other participation categories. 'Non-participator' households have much smaller families, have less education and older household heads, participate less in CBOs, and have much less access to extension services and market information. 'Non-participators' seem very unlikely to participate without similar having conditions as 'participators.' Second, the multinomial regression analysis did not reveal many distinguishing characteristics between continuous 'participators' and 'dis-participators,' especially among production related IAA value chain actors. The number of household assets was the only significant distinguishing characteristic between production activity 'participators' and 'dis-participators,' while household head age, the number of assets, farm size, fisheries income, and non-farm income were significant distinguishing characteristics between these two groups among up and down stream (non-production) value chain actors. Thus, households with younger heads, fewer assets, larger farms, less fisheries income and higher non-farm income are more likely to discontinue participation in up and down stream IAA value chain activities. Yet, current rates of CBO membership and access to extension services and market information were actually quite high among 'dis-participators' (e.g. over 75% of up and downstream value chain activity 'dis-participators' have access to market information), which suggests that while they may be critical determinants of IAA value chain participation, these does not guarantee that such participation will be profitable. Thus, changes in access to these complementary factors do not appear to be likely to make a significant difference in terms of transitioning 'dis-participators' back to IAA participation.

Resource poor farmers are capable of participating in IAA value chains, especially in up and downstream value chain activities. Controlling for fixed and random omitted variables and potential endogenous regressors revealed that farm size is not a key determinant of participation in either production or non-production

IAA value chain activities despite often being featured in technology adoption literature. This is especially possible for up and downstream value chain activity participants because land ownership is not necessary to perform these activities and small properties or rented land will suffice for participating in IAA production value chain activities. This result is robust to alternative specifications and estimation techniques. The probability of participating in IAA value chain activities increases significantly with education, and together with greater CBO membership and access to extension services and market information they confirm the broader positive effects of technical knowledge and human capital development on value chain participation.

The analyses featured in Chapter 4 investigated the relationship between IAA value chain participation dynamics and household economic wellbeing in order to examine whether or not IAA is a sustainable agricultural intensification option that contributes to poverty reduction and food security in developing countries, especially in Asia which is home to the largest populations that face poverty and food security challenges. We estimated the welfare impacts of IAA participation dynamics through different specifications (e.g. POLS, RE, FE, Heckit and Control function approaches) under different assumptions to control for unobserved heterogeneity and endogenous selection of IAA value chain participation dynamics. Beginning with a naive POLS estimation that assumes that participation and dis-participation in IAA value chains are exogenous, we further controlled for unobserved heterogeneity and endogenous selection. The results validated as well as added to the current understanding of technology adoption. Additionally, we applied FE specifications to sample household disaggregated by production related actors (that participated in production related IAA value chain activities that require land) and the non-production value chain actors (extremely poor households, most of which do not have access to land), to explore the distribution of IAA value chain participation benefits among all IAA value chain actors.

The results of the IAA value chain participation dynamics impacts analyses are robust across specifications, thereby justifying concerns about unobserved heterogeneity and endogenous selection with respect to IAA value chain participation. There is consistent evidence of a positive relationship between IAA value chain participation and household income and the consumption frequency of important food items such as fish and pulse. The results also indicate that these positive income effects increased over time and that IAA value chain participation benefits were comparatively higher for the relatively wealthier households that participated in production related value chain activities. Considering participation dynamics, abandonment of IAA value chain participation (dis-participation) impacts negatively to household income, which suggests that the decision to discontinue IAA participation is not based on the economic superiority of alternative options, but rather may be due to other IAA value chain participation barriers. Based on the results of our examination of IAA value chain participation dynamics, participation appears to improve the welfare of poor and marginalized indigenous households in Bangladesh.

In the analyses presented in Chapter 5 we argued that integrated rice–fish systems provide a more sustainable alternative to conventional rice monoculture systems. We investigated the comparative impacts of integrated rice–fish systems and rice monoculture systems on socio-environmental variables by exploring farm level environmental input use and farmer perceptions, and an extensive literature review. To date there do not appear to have been any attempts to examine the socio-environmental performance of integrated rice–fish production relative to rice monoculture based on farmer perceptions. We also examined the determinants of farmer perceptions about the socio-environmental impacts of rice monoculture. We investigated this question empirically using two econometric models that correct for the presence of selection bias that influences on integrated rice–fish system adoption and consequently on farmer perceptions.

Stepwise, we first compared farm level input use between integrated rice–fish systems and rice monoculture (Table 5.2) and reviewed related literature. The results indicate that integrated rice–fish farmers use less environmentally hazardous inputs and more environmentally benign inputs than rice monoculture farmers. The results indicate that rice–fish systems not only have less negative environment impacts, but also performed better with respect to socio-economic indicators. The literature review results also support and validate the farm level analysis results.

Second, we investigated farmer perceptions on the adverse socio-environmental impacts of both farming systems. Integrated rice–fish farmers were asked about both types of farming systems, however, rice monoculture farmers were only asked about the impacts of rice monoculture system because they lack direct experience with integrated rice–fish systems. From the results it is evident that both types of farmers were well aware of many of the positive and negative socio-environmental impacts of the respective production systems. Integrated rice–fish farmers perceived this system to have less negative socio-environmental impacts than rice monoculture (Table 5.3). The adverse socio-environmental impacts of rice monoculture were perceived by both types of farmers, but integrated rice–fish farmers were more aware of these negative impacts than rice monoculture farmers (Table 5.4). All of these results are based on perception indices from farmer responses to questions about different socio-environmental indicators and the perceived severity of identified impacts. In most cases farmer perceptions remain limited to observable impacts such as changes in diseases and pests, soil characteristics, the relative quantity of beneficial organisms (e.g. earthworms, frogs, etc.), and fish harvests. Thus the socio-environmental perceptions of farmers and secondary evidence supported the farm level comparison results (Table 5.2).

We examined the factors that influence farmer perceptions about the socio-environmental impacts of rice monoculture using a Tobit and PSM approach. The Tobit results show that ethnicity (indigenous or non-indigenous), off-farm income, AFP participation, adoption of integrated rice–fish production, and infrastructure access were the most significant determinants of socio-environmental impacts perceptions (Table 5.5). The PSM results further supported the robustness of integrated rice–fish system adoption as a major determinant of farmer awareness of the adverse socio-environmental impacts of rice monoculture (Table 5.6). Thus promotion of integrated rice–fish systems, development of better infrastructure, and relevant education through FFS approaches would be expected to enhance socio-environmental awareness and may help reduce the use of environmental hazardous inputs and improve the adoption rates of integrated rice–fish systems.

6.2 Conclusions and Policy Implications

The overall research goal was to assess IAA value chain participation dynamics and impacts in Bangladesh based on consideration of all IAA value chain participating actors and non-participants. We applied value chain, partial budgeting and SWOT analyses as evaluation tools to assess the economic performance and identify the key factors that affect the adoption and diffusion of rice–fish based IAA among indigenous farmers in Bangladesh. Value chain mapping showed that there is little processing of harvested fish (mainly icing, grading and transportation) and that only a short period is required from harvest to final consumption (typically the same day) due to the live/fresh nature of all sales in the relatively short chain. Overall the quantitative results of the gross margin, partial budgeting and gender disaggregated employment analyses reveal positive benefits from the replacement of rice monoculture with integrated rice–fish systems. Indeed the enormous opportunities for further improvement to rice–fish based IAA and associated value-chain performance provides a strong argument for engagement in these chains by the private sector and for greater government support in the form of policy and legislation (on issues such as land tenure rights, access to credit and markets, the quality of and access to irrigation water and feed, the development of related infrastructure, and public and private sector human capacity development and training).

The IAA value chain participation dynamics analysis results confirm some previous findings. More educated household heads, larger families, CBO membership, and better access to extension services and market information were associated with participation and continued participation in IAA value chain activities. Importantly, farm size and farm income did not appear to be significant positive determinants of continuous IAA value chain participation, suggesting that IAA value chain activities are appropriate low-input activities for resource-poor households. The results also indicate that the factors contributing to continuous participation and discontinued IAA value chain participation vary, especially for up and downstream activities. Important determinants of disparticipation were household head age, the number of assets, farm size, fisheries income, and non-farm income. As household head education, family size (as a proxy for family labour availability), CBO membership, and access to extension services and market information are important to IAA value chain participation; policy efforts to strengthen technical knowledge among farmers should provide platforms for active interaction among stakeholders as argued by innovation systems theory. A good example is the effective use of broad extension services and research networks, the FFS approach, and existing CBOs, which are intended to facilitate interactions among different actors. The findings of this study suggest that policy that promotes IAA adoption and diffusion should focus on education, institutional and technological innovations that facilitate interactions between participants and non-participants, and provide the training and opportunities to develop the necessary technical knowledge and practical work skills.

Participation in IAA value chains was positively correlated with household income, and consumption, especially fish consumption. The benefits of participation do not continue to accrue after participation is discontinued. We found evidence that IAA value chain participation had higher impacts on the welfare of relatively wealthier households that participated in production related value chain activities relative to extremely poor landless households that participated in upstream and downstream IAA value chain activities. The results highlight the importance of IAA value chain participation for poor smallholders and how participation may contribute to food security and poverty reduction among rural smallholders. Cost effective agricultural policies that help to create an enabling environment for sustainable technology adoption and continuation can contribute significantly to improved food security and poverty reduction in rural areas.

The findings of this study suggest that rice–fish based IAA can be a more sustainable agricultural practice than rice monoculture. Plot level data on inputs and outputs and farmer perceptions indicate that rice–fish based IAA performs better than monoculture with respect to socio-economical and environmental criteria. Although farmers are well aware of the more negative socio-environmental impacts of rice monoculture relative to integrated rice–fish systems, their perceptions are limited to observable impacts. The promotion of the FFS or similar institutional approaches and infrastructure development would likely have positive roles in raising awareness about the adverse socio-environmental impacts of rice monoculture.

Overall the results of the study indicate the need for broader attention to IAA. Although the results generally reveal that the gains from IAA value chain participation are substantial, many smallholders are unable to participate or cannot sustain participation due to various factors. Thus policy and institutional interventions are necessary to stimulate the adoption and diffusion of IAA among poor smallholders as a means of improving nutrition, incomes, and ameliorating poverty.

6.3 Further Research Needs

Although this study is the first of its kind with respect to consideration of several aspects of technology adoption and its impacts, there are several issues that were not considered or were beyond the scope of this study that should be addressed by future research efforts. Although the results show that rice–fish based IAA systems create employment opportunities that can be occupied by women, the resulting effects on intra-household labour allocation and reproductive roles (e.g. childbearing, child care) need further study. The value chain analysis was limited to a few actors along the IAA value chain and that were engaged in rice monoculture. Future research efforts should consider all actors involved in IAA and rice monoculture value chains to provide more comprehensive and comparative results.

To better understand the IAA value chain adoption/participation dynamics, plot/farm level ecological characteristics, farmer attitudes regarding risk, and farmer perceptions about new technology are important factors for consideration, especially for participation in value chain production processes. Future research efforts that consider these factors will provide insight that could be compared to the findings of this study. There are increasing attempts to promote more sustainable agricultural innovation and intensification options, and the findings of this study imply some possible pathways. These results also have significant implications for the dissemination and legitimacy of sustainable innovation and intensification options. Additional research on the welfare impacts of these innovations on farmers are needed to further strengthen arguments in support of more sustainable innovations and intensification.

We only used household income and consumption data in the analyses. Future research using alternative welfare indicators such as asset indices that consider their monetary value would be helpful to better reveal the relationships between participation and household capital. Dietary diversity indices and gender disaggregated analyses will also provide a clearer understanding of food security and gender implications of IAA adoption. In addition, the adoption of new 'technologies,' much like IAA value chain participation, not only has direct impacts, but may also have indirect impacts (e.g. spill over effects) that were beyond the scope of this study. Thus future research that takes into account economy-wide effects using appropriate economy wide modelling approaches (see Subramanian and Qaim, 2009) would provide a better understanding of the broader impacts of IAA.

Finally, the variables or indicators that we used to assess the relative environmental impacts of rice monoculture and rice–fish based IAA systems are not hard indicators (i.e. direct measurements of chemical or physical indicators of soil health, water quality, public health, etc.). Rather we have tried to address these issues by using different perception based indicators and plot level comparisons of the environmental impacts of input use, and robust estimation techniques. Future research application within a conceptual framework that more fully accounts for those considerations (i.e. hard environmental indicators) will provide results that would complement the findings of this study. The research findings indicate that rice–fish based IAA adoption is a robust determinant of farmer perceptions on the socio-environmental impacts of rice-monoculture, but due to data limitations we could not reveal whether or not this heightened perception leads to lesser use of environmental hazardous inputs or greater use of environmentally benign inputs.

References

Abadie, A., Drukker, D., Leber Herr, J., Imbens, G. W. (2004). Implementing matching estimators for average treatment effects in Stata. *Stata Journal* 4 (3), 290–311.

Abdullah, A. R., Bajet, C. M., Matin, M. A., Nhan, D. D., and Sulaiman, A. H. (1997). Ecotoxicology of pesticides in the tropical paddy field ecosystem. *Environmental toxicology and chemistry*, 16(1), 59–70.

ADB (2014a). Key Indicators for Asia and the Pacific 2014. Special chapter: Poverty in Asia: A Deeper Look. Philippines: Asian Development Bank. http://www.adb.org/sites/default/files/pub/2014/ki2014.pdf.

ADB (2014b). Bangladesh quarterly economic update (QEU). Bangladesh resident mission, Asian Development Bank, Dhaka, Bangladesh. http://www.adb.org/sites/default/files/ban-qeu-2014-03.pdf.

ADB (Asian Development Bank), (2005). An Overview of Small-Scale Freshwater Aquaculture in Bangladesh. In: Special Evaluation Study on Small-scale Freshwater Rural Aquaculture Development for Poverty Reduction—Case Studies. ADB, Manila.

Adelman, I., Taylor, J. E., and Vogel, S. (1988). Life in a Mexican village: a SAM perspective. *The Journal of Development Studies*, 25(1), 5–24.

Adesina, A. A., and Baidu-Forson, J. (1995). Farmers' perceptions and adoption of new agricultural technology: evidence from analysis in Burkina Faso and Guinea, West Africa. *Agricultural economics*, 13(1), 1–9.

Adesina, A. A., and Zinnah, M. M. (1993). Technology characteristics, farmers' perceptions and adoption decisions: A Tobit model application in Sierra Leone. *Agricultural economics*, 9(4), 297–311.

AFP (Adivasi Fisheries Project), (2010). Fisheries and Aquaculture Enterprise Development for Adivasi (Tribal) Communities in the Northern and Northwestern Regions of Bangladesh. The European Union's Food Security Programme for Bangladesh. WorldFish Center Bangladesh and South Asia office, Dhaka, Bangladesh.

Ahmed, A. U., Ahmad, K., Chou, V., Hernandez, R., Menon, P., Naeem, F., Naher, F., Quabili, W., Sraboni, E., Yu, B., and Hassan, Z. (2013). The Status of Food Security in the Feed the Future Zone and Other Regions of Bangladesh: Results from the 2011–2012 Bangladesh Integrated Household Survey. Project report submitted to the US Agency for International Development. International Food Policy Research Institute, Dhaka. http://www.ifpri.org/sites/default/files/publications/bihstr.pdf.

Ahmed, A. U., Ruth Vargas Hill, Lisa C. Smith, Doris M. Wiesmann, and Tim Frankenberger (2007). The World's deprived: Characteristics and causes of extreme poverty and hunger. 2020 discussion paper 43 "A 2020 Vision for Food, Agriculture, and the Environment" International Food Policy Research Institute Washington, DC October 2007. http://ageconsearch.umn.edu/bitstream/42252/2/vp43.pdf (accessed on 15.10.2011).

Ahmed, M. (1992). Status and potential of aquaculture in aquaculture in small waterbodies (ponds and ditches) in Bangladesh. ICLARM Tech. Rep. 37, 36p.

Ahmed, N., and Garnett, S. T. (2010). Sustainability of freshwater prawn farming in rice fields in southwest Bangladesh. *Journal of Sustainable Agriculture*, 34(6), 659–679.

Ahmed, N., and Garnett, S. T. (2011). Integrated rice–fish farming in Bangladesh: meeting the challenges of food security. *Food Security*, 3(1), 81–92.

Ahmed, N., Demaine, H., and Muir, J. F. (2008). Freshwater prawn farming in Bangladesh: history, present status and future prospects. *Aquaculture Research*, 39(8), 806–819.

Ahmed, N., Wahab, M. A., and Thilsted, S. H. (2007). Integrated aquaculture-agriculture systems in Bangladesh: potential for sustainable livelihoods and nutritional security of the rural poor. *Aquaculture Asia*, 12(1), 14–22.

Ahmed, N., Ward, J. D., and Saint, C. P. (2014). Can integrated aquaculture-agriculture (IAA) produce "more crop per drop"? *Food Security*, 6(6), 767–779.

Ahmed, N., Zander, K. K., and Garnett, S. T. (2011). Socioeconomic aspects of rice-fish farming in Bangladesh: opportunities, challenges and production efficiency. *Australian Journal of Agricultural and Resource Economics*, 55(2), 199–219.

Alam, M. F., Thomson, K. J. (2001). Current constraints and future possibilities for Bangladesh fisheries. *Food policy*, 26(3), 297–313.

Alauddin, M., and Quiggin, J. (2008). Agricultural intensification, irrigation and the environment in South Asia: Issues and policy options. *Ecological Economics*, 65(1), 111–124.

Alauddin, M., and Tisdell, C. A. (1998). *The environment and economic development in South Asia: an overview concentrating on Bangladesh*. London: Macmillan Press Ltd.

Alauddin, M., Tisdell, C. (1991). *The "Green Revolution" and economic development: the process and its impact in Bangladesh*, London: Macmillan Press Ltd.

Alem, Y., Hassen, S., and Köhlin, G. (2013). The Dynamics of Electric Cookstove Adoption: Panel data evidence from Ethiopia. Environment for Development: Discussion Paper Series, EfDDP, 13–30. http://rff.org/RFF/Documents/EfD-DP-13-03.pdf.

Ali, M. A., Yasmin, S., Hamid, M. A., and Islam, M. A. (1995). Effect of tilapia stocking on the growth of carps in integrated duck-cum-fish farming ponds. *Bangladesh Journal Fisheries, 15*, 1–7.

Ali, M. H., and Mateo, L. G. (2007). Economics of rice–fish culture in wet land rice ecosystems. *SAARC Journal of Agriculture, 5*(2), 1–6.

Ali, M. M., Saheed, S. M., Kubota, D., Masunaga, T., and Wakatsuki, T. (1997). Soil degradation during the period 1967–1995 in Bangladesh: II. selected chemical characters. *Soil Science and Plant Nutrition, 43*(4), 879–890.

Alsagoff, S. A. K., Clonts, H. A., and Jolly, C. M. (1990). An integrated poultry, multi-species aquaculture for Malaysian rice farmers: a mixed integer programming approach. *Agricultural Systems, 32*(3), 207–231.

Alston, J. M., and Pardey, P. G. (2001). Attribution and other problems in assessing the returns to agricultural R&D. *Agricultural Economics, 25*(2–3), 141–152.

Amsalu, A., and De Graaff, J. (2007). Determinants of adoption and continued use of stone terraces for soil and water conservation in an Ethiopian highland watershed. *Ecological Economics, 61*(2), 294–302.

An, H. (2008). The adoption and disadoption of recombinant bovine somatotropin in the US dairy industry. In American Agricultural Economics Association Annual Meeting, Orlando, FL. http://ageconsearch.umn.edu/bitstream/6278/2/465503.pdf.

Angrist, J. D. (2001). Estimation of limited dependent variable models with dummy endogenous regressors. *Journal of Business and Economic Statistics, 19*(1), 2–28.

Antle, J. M., and Pingali, P. L. (1994). Pesticides, productivity, and farmer health: A Philippine case study. *American Journal of Agricultural Economics, 76*(3), 418–430.

Arce, R. G. and Dela Cruz, C. R. (1979). Yield trials on rice–fish culture at the Freshwater Aquaculture Center. *Fisheries Research Journal of the Philippines, 4*, 1–8.

Arevalo, T. Z., (1987). The rice–fish culture program. Paper presented at the fisheries forum on "The Development in Integrated Agri-Aquaculture Farming systems." Bureau of Fisheries and Aquatic Resources, Quezon City, Philippines, 12 pp.

Azim, M. E., Wahab, M. A., (1998). Effects of duckweed (Lemna sp.) on pond ecology and fish production in carp polyculture of Bangladesh. *Bangladesh Journal of Fisheries, 21*, 17–28.

Badiuzzaman, M., Cameron, J., and Murshed, S. M. (2013). Livelihood decisions under the shadow of conflict in the Chittagong Hill Tracts of Bangladesh. WIDER Working Paper. No. 2013/006). http://www.wider.unu.edu/publications/

working-papers/2013/en_GB/wp2013-006/_files/89087242609754248/default/wp2013-006.pdf.

Baecke, E, Rogiers G, De. Cock, L., Van Huylenbroeck, G. (2002). The supply chain and conversion to organic farming in Belgium or the story of the egg and the chicken. *British Food Journal*, 104 (3/4/5), 163–74.

Baffes, J., and Gautam, M. (2001). Assessing the sustainability of rice production growth in Bangladesh. *Food Policy*, 26(5), 515–542.

Bahr, M, Botschen, M, Laberentz, H, Naspetti, S, Thelen, E, Zanoli R (2004). The European Consumer and Organic Food. School of Management and Business, University of Wales, Aberystwyth.

Baidu-Forson, J. (1999). Factors influencing adoption of land-enhancing technology in the Sahel: lessons from a case study in Niger. *Agricultural Economics*, 20(3), 231–239.

Banglapedia. (2014). National Encyclopedia of Bangladesh. http://en.banglapedia.org/index.php?title=Main_Page.

Barham, B. L. (1996). Adoption of a politicized technology: bST and Wisconsin dairy farmers. *American Journal of Agricultural Economics*, 78(4), 1056–1063.

Barham, B. L., Foltz, J. D., Jackson-Smith, D., and Moon, S. (2004). The dynamics of agricultural biotechnology adoption: Lessons from series rBST use in Wisconsin, 1994–2001. *American Journal of Agricultural Economics*, 86(1), 61–72.

Barkat, A., Hoque, M., Halim, S., Osman, A., (2009). *Life and Land of Adivasis: Land Dispossession and Alienation of Adivasis in Plain Districts of Bangladesh*. Dhaka: Pathak Shamabesh.

Barnard, C. S, Nix, J. J. (1979). *Farm Planning and Control*. Cambridge: Cambridge University Press, 600 pp.

Barrientos S, Dolan C, Tallontire A. (2003). A gendered value chain approach to codes of conduct in African horticulture. *World Development*, 31(9), 1511–1526.

Barman, B. K., and Little, D. C. (2006). Nile tilapia (Oreochromis niloticus) seed production in irrigated rice-fields in Northwest Bangladesh—an approach appropriate for poorer farmers? *Aquaculture*, 261(1), 72–79.

BBS, (2011). Household Income and Expenditure Survey-2010. Dhaka: Bangladesh Bureau of Statistics, Ministry of Planning, Government of the People's Republic of Bangladesh.

BBS. (2013). Statistical Yearbook of Bangladesh. Statistics and Informatics Division. Ministry of Planning.

BBS. (2012). Statistics and Information Division, Ministry of Planning, Government of the People's Republic of Bangladesh. Bangladesh Population and Housing Census 2011: Socio-Economic and Demographic Report. National Report Volume 4.

Becerril, J., Abdulai, A. (2010). The impact of improved maize varieties on poverty in Mexico: a propensity score-matching approach. *World Development*, 38(7), 1024–1035.

Becker, S. O., Ichino, A., (2002). Estimation of average treatment effects based on propensity scores. *The Stata Journal, 2*(4), 358–377.

Belton, B., and Azad, A. (2012). The characteristics and status of pond aquaculture in Bangladesh. *Aquaculture, 358*, 196–204.

BER. (2012). Bangladesh Economic Review. Economic Adviser's Wing, Finance Division, Ministry of Finance, Government of the People's Republic of Bangladesh.

BER, (2014). Bangladesh Economic Review. Economic Adviser's Wing, Finance Division, Ministry of Finance, Government of the People's Republic of Bangladesh, Dhaka, Bangladesh.

Bera, A. K., and Kelley, T. G. (1990). Adoption of high yielding rice varieties in Bangladesh: an econometric analysis. *Journal of Development Economics, 33*(2), 263–285.

Berg, H. (2001). Pesticide use in rice and rice–fish farms in the Mekong Delta, Vietnam. *Crop Protection, 20*(10), 897–905.

Berg, H. (2002). Rice monoculture and integrated rice–fish farming in the Mekong Delta, Vietnam—economic and ecological considerations. *Ecological Economics, 41*(1), 95–107.

Berhane, G., and Gardebroek, C. (2011). Does microfinance reduce rural poverty? Evidence based on household panel data from northern Ethiopia. *American Journal of Agricultural Economics*, 93 (1), 43–55.

Bernard, H. (2006). *Research methods in anthropology: qualitative and quantitative approaches.* Oxford, United Kingdom: Alta Mira Press.

Besley, T., and Case, A. (1993). Modeling technology adoption in developing countries. *The American Economic Review*, 83 (2), 396–402.

Bester, A., and Hansen, C. (2007). "Flexible correlated random effects estimation in panel models with unobserved heterogeneity," Technical report, mimeo. http://citeseerx.ist.psu.edu/viewdoc/download?doi=10.1.1.297.1249&rep=rep1&type=pdf.

Bezu, S., Kassie, G. T., Shiferaw, B., and Ricker-Gilbert, J. (2014). Impact of improved maize adoption on welfare of farm households in Malawi: a panel data analysis. *World Development, 59*, 120–131.

Bhuiyan, N. I., Paul, D. N. R., and Jabber, M. A. (2002). Feeding the extra millions by 2025 – Challenges for rice research and extension in Bangladesh. National Workshop on Rice Research and Extension in Bangladesh, Bangladesh Rice Research Institute, Gazipur, 29–31 January.

Blythe, J. L. (2013). Social-ecological analysis of integrated agriculture-aquaculture systems in Dedza, Malawi. *Environment, development and sustainability, 15*(4), 1143–1155.

Bolwig, S, Ponte, S, Du Toit, A, Riisgaard, L, Halberg, N. (2010). Integrating poverty and environmental concerns into value-chain analysis: a conceptual framework. *Development Policy Review, 28*(2), 173–194.

Bosma, R. H., Nhan, D. K., Udo, H. M., and Kaymak, U. (2012). Factors affecting farmers' adoption of integrated rice–fish farming systems in the Mekong delta, Vietnam. *Reviews in Aquaculture, 4*(3), 178–190.

Bottrell, D. G., and Weil, R. R. (1995). Protecting crops and the environment: striving for durability. In S. R. Juo, and R. D. Freed eds. Agriculture and Environment: Bridging Food Production and Environmental Protection in Developing Countries. ASA Special Publication No. 60. Madison, WI: American Society of Agronomy, Crop Science Society of America, Soil Science Society of America, 55–73.

Bravo-Ureta, B. E., Solis, D., Cocchi, H., and Quiroga, R. E. (2006). The impact of soil conservation and output diversification on farm income in Central American hillside farming. *Agricultural Economics, 35*(3), 267–276.

BRKB (2010). Rice Statistics in Bangladesh. Bangladesh rice knowledge bank (BRKB), Bangladesh Rice Research Institute, Gazipur, Bangladesh. http://www.knowledgebank-brri.org/riceinban.php (accessed on 10.01.2015).

Bromley, D. W. (2010). Food security: beyond technology. *Science, 328*(5975), 169–169.

Brown, L. R. (1988). The changing world food prospect: the nineties and beyond. Worldwatch Paper No. 85. Worldwatch Institute, Washington, DC.

Brummett, R. E. (1999). Integrated aquaculture in subsaharan Africa. *Environment, Development and Sustainability, 1*(3–4), 315–321.

Brummett, R. E., Noble, R. P., (1995). Aquaculture for African smallholders. ICLARM Tech. Rep. 46. WorldFish Center, Penang, Malaysia, 69.

Bryan, G., Chowdhury, S., and Mobarak, A. M. (2014). Underinvestment in a Profitable Technology: The Case of Seasonal Migration in Bangladesh. *Econometrica, 82*(5), 1671–1748.

Buck, O. H., R. J. Baur, and S. R. Rose. (1979). Experiments in recycling swine manure in fishponds. pp. 489–492. in Pillay, T. V. R. and W. A. Dill. (ed.). Advances in Aquaculture, Fishing. News Book, London.

Cagauan, A. G. (1995). Overview of the potential roles of pisciculture on pest and disease control and nutrient management in rice fields. In The Management of Integrated Freshwater Agro-Piscicultural Ecosystems in Tropical Areas (Eds J. J. Symoens & J. C. Micha), pp. 203–244. Brussels, CTA and Royal Academy of Overseas Sciences.

Cagauan, A. G., Dela Cruz, C. R., Florblanco, F., Cruz, E. M. and Sevilleja, R. C. (1994a). Impacts of fish and pesticides. In Role of Fish in Enhancing Ricefield Ecology and in Integrated Pest Management (Ed. C. R. dela Cruz), pp. 22–23. ICLARM Conference Proceedings 43. Manila: ICLARM.

Cagauan, A. G., Dela Cruz, C. R. and Lightfoot, C. (1994b). Nitrogen models of lowland irrigated ecosystems with and without fish using ECOPATH II. In Role of Fish in Enhancing Ricefield Ecology and in Integrated Pest Management (Ed. C. R. dela Cruz), p. 36. ICLARM Conference Proceedings 43. Manila: ICLARM.

Cagauan, A., and R. Arce. (1992). Overview of Pesticide Use in Rice–Fish Farming in Southeast Asia. In C. R. dela Cruz, C. Lightfoot, B. A. Costa-Pierce, et al. (eds.). Rice–fish Research and Development in Asia. ICLARM Conference Proceedings 24, pp. 217–33. http://www.worldfishcenter.org/resource_centre/WF_102.pdf (accessed on 08.01.2015).

Cagauan, A. G. (1999). Production, economics and ecological effects of Nile tilapia (Oreochromis niloticus L.), a hybrid aquatic fern azolla (Azolla microphylla Kaulf.? Azolla filiculoides Lam.) and mallard duck (Anas platyrhynchos L.) in integrated lowland irrigated rice-based farming systems in the Philippines. Ph.D. thesis, Universite Catholique de Louvain.

Caliendo, M., and Kopeinig, S. (2008). Some practical guidance for the implementation of propensity score matching. *Journal of economic surveys*, 22(1), 31–72.

Cameron, L. A. (1999). The importance of learning in the adoption of high-yielding variety seeds. *American Journal of Agricultural Economics*, 81(1), 83–94.

Carletto, C., de Janvry, A., Sadoulet, E., (1996). Knowledge, Toxicity, and Internal Shocks: The Determinants of Adoption and Abandonment of Non-traditional Export Crops by Smallholders in Guatemala (Working Paper No. 791). Department of Agricultural and Resource Economics, University of California, Berkeley. http://ageconsearch.umn.edu/bitstream/25088/1/wp791.pdf.

Chamberlain, G. (1984). "Panel Data", in Zvi Griliches and Michael Intriligator (eds.), Handbook of Econometrics, Amsterdam: North-Holland.

Chen, S., and Ravallion, M. (2012). More relatively-poor people in a less absolutely-poor world. *Policy Research Working Paper*, 6114. http://elibrary.worldbank.org/doi/pdf/10.1596/1813-9450-6114.

Chernozhukov, V., J. Hahn, and Newey, W. (2005). Bound analysis in panel models with correlated random effects. Technical report, MIT, UCLA.

Chowdhury, M. R. (2009). *Population challenge facing Bangladesh*. CW Post Campus, New York: Long Island University.

Christensen, V, Steenbeek, J, Failler, P. (2011). A combined ecosystem and value chain modeling approach for evaluating societal cost and benefit of fishing. *Ecological Modelling*, 222(3), 857–864.

Coche, A. G. (1967). Fish culture in rice fields a world-wide synthesis. *Hydrobiologia*, 30(1), 1–44.

Coelli, T., Rahman, S., and Thirtle, C. (2003). A stochastic frontier approach to total factor productivity measurement in Bangladesh crop agriculture, 1961–92. *Journal of International Development*, 15(3), 321–333.

Coles, C., Mitchell, J. (2010). Gender and agricultural value chains: a review of current knowledge and practice and their policy implications. ESA Working Paper No. 11–05. Rome, Italy: Agricultural Development Economics Division. FAO SOFA. 1–29. <http://www.fao.org/docrep/013/am310e/am310e00.pdf> (accessed on 26.08.2013).

Conley, T., and Udry, C. (2001). Social learning through networks: The adoption of new agricultural technologies in Ghana. *American Journal of Agricultural Economics*, 83 (3), 668–673.

Conway, G. (1999). *The doubly green revolution: food for all in the twenty-first century*. Ithaca, NY: Cornell University Press.

Conway, G. (2011). The doubly green revolution. In Becker et al., (eds) Development on the margin. Tropentag 2011. International Research on Food Security, Natural Resource Management and Rural Development. http://dpg.phytomedizin.org/fileadmin/tagungen/07_Tropentag/Doku/Tropentag%202011.pdf.

Cortijo, M. J. A. (2014). Contributing to the eradication of hunger, food insecurity and malnutrition: lessons from Bangladesh. ESA Working Paper No. 14–06. Rome, FAO. http://www.fao.org/3/a-i3868e.pdf.

Costa-Pierce, B. A., (2002). Ecology as the paradigm for the future of aquaculture. In: Costa-Pierce, B. A. (Ed.), Ecology Aquaculture–The Evolution of the Blue Revolution. Blackwell Science, pp. 339–372.

Courtney P, Mayfield L, Tranter R, Jones P, Errington A, (2007). Small towns as "sub-poles" in English rural development: Investigating rural–urban linkages using sub-regional social accounting matrices. *Geoforum*, 38 (6), 1219–1232.

Crost, B., Shankar, B., Bennett, R., and Morse, S. (2007). Bias from farmer self-selection in genetically modified crop productivity estimates: evidence from Indian data. *Journal of Agricultural Economics*, 58(1), 24–36.

Dalsgaard, J. P. T., Oficial, R. T., (1997). A quantitative approach for assessing the productive performance and ecological contributions of smallholder farms. *Agricultural Systems* 55 (4), 503–533.

Dalsgaard, J. P. T., Prein, M., (1999). Integrated smallholder agriculture-aquaculture in Asia: optimizing trophic flows. In: Smaling, E. M. A.; O. Oenema;

L. O. Fresco (Eds.) *Nutrient Disequilibria in Agro-ecosystems. Concept and Case Studies*. CABI Publishing. CAB International, Wallingford, Oxon, pp. 141–155.

Das, D. R., Qaddus, M. A., Khan, A. H., and Nur-e-Elahi, M. (2002). Farmers participatory productivity evaluation of integrated rice and fish systems in transplanted *aman* rice. *Pakistan Journal of Agronomy*, 1(2–3), 105–106.

Das, R. (2002). The green revolution and poverty: a theoretical and empirical examination of the relation between technology and society. *Geoforum*, 33 (1): 55–72.

Dashu, N., Yinghong, C., Jianguo, W., (1992). Mutualism of rice and fish in ricefields. In: dela Cruz, C. R., Lightfoot, C., Costa-Pierce, B. A., Carangal, V. R., Bimbao, M. P. (Eds.), *Rice–fish research and development in Asia*. International Center for Living Aquatic Resources Management, Manila, Philippines, pp. 173–175. http://www.worldfishcenter.org/resource_centre/WF_102.pdf (accessed on 08.01.2015).

Datta, A., Nayak, D. R., Sinhababu, D. P., and Adhya, T. K. (2009). Methane and nitrous oxide emissions from an integrated rainfed rice–fish farming system of Eastern India. *Agriculture, ecosystems and environment*, 129(1), 228–237.

Dela Cruz, C. R. (Ed.), (1994). Role of fish in enhancing ricefield ecology and integrated pest management. ICLARM Conference Proceeding 43, International Center for Living Aquatic Resources Management, Manila, 50 pp. http://www.worldfishcenter.org/libinfo/Pdf/Pub%20CP6%2043.pdf (accessed on 12-12-2014).

Devendra, C., and Thomas, D. (2002). Smallholder farming systems in Asia. *Agricultural Systems*, 71(1), 17–25.

Dewan, S., (1992). Rice-fish farming systems in Bangladesh: past, present and future. In: dela Cruz, C. R., Lightfoot, C., Costa-pierce, B. A., Carangal, V. R., Bimbao, M. P. (Eds.), ICLARM Conference Proceedings of the Rice–Fish Research and Development in Asia, vol. 24, pp. 457.

Dey, M. M., Paraguas, F. J., Kambewa, P., and Pemsl, D. E. (2010). The impact of integrated aquaculture–agriculture on small-scale farms in Southern Malawi. *Agricultural Economics*, 41(1), 67–79.

Dey, M. M., Prein, M., Mahfuzul Haque, A. B. M., Sultana, P., Cong Dan, N., and Van Hao, N. (2005). Economic feasibility of community-based fish culture in seasonally flooded rice fields in Bangladesh and Vietnam. *Aquaculture Economics and Management*, 9(1–2), 65–88.

Dey, M. M., Spielman, D. J., Haque, A. B. M. M., Rahman, M. S., and Valmonte-Santos, R. (2013). Change and Diversity in Smallholder Rice–fish Systems: Recent Evidence and Policy Lessons from Bangladesh. *Food Policy*, 43, 108–117.

Dey, M. M., Kambewa, P., Prein, M., Jamu, D., Paraguas, F. J., Briones, R. M., (2007). World fish center: Impact of the development and dissemination of integrated aquaculture-agriculture technologies in Malawi. In: Waibel, H., Zilberman, D. (Eds.), International Research on Natural Resource Management: Advances in Impact Assessment. FAO and CABI, Wallingford, pp. 147–168.

Dey, M. M., Prein, M., (2006). Community based fish culture in seasonal floodplains. *NAGA, WorldFish Center Quarterly*, 29 (1&2), 21–27. http://www.worldfishcenter.org/resource_centre/community.pdf.

Dey, NC and Haq, F. (2009). Study of the impact of intensive cropping on the long term degradation of natural resources in some selected agroecological regions of Bangladesh. Center for Agri-research and Sustainable Environment and Entrepreneurship Development (CASEED). http://www.nfpcsp.org/agridrupal/sites/default/files/Final_Report_CF2_Approved.pdf.

Dillon, J. L, Hardaker, J. B. (1989). Farm management research for small farmer development (Vol. 41). Food & Agriculture Organization (FAO), Rome, Italy.

Dinar, A., and Yaron, D. (1992). Adoption and abandonment of irrigation technologies. *Agricultural economics*, 6(4), 315–332.

DOF. (2010). Fisheries Statistical Yearbook of Bangladesh. Department of Fisheries, Ministry of Fisheries and Livestock, Dhaka, Bangladesh.

Dolan, C., Humphrey, J. (2004). Changing governance patterns in the trade in fresh vegetables between Africa and the United Kingdom. *Environment and planning*, 36(3), 491–510.

Doss, C. R. (2006). Analyzing technology adoption using microstudies: limitations, challenges, and opportunities for improvement. *Agricultural Economics*, 34(3), 207–219.

DSAP (Development of Sustainable Aquaculture Project). (2005). Final report of the Development of Sustainable Aquaculture Project (DSAP). Prepared by the World-Fish Center, Bangladesh and South Asia Office, Dhaka, Bangladesh.

Dugan, P., Dey, M. M., and Sugunan, V. V. (2006). Fisheries and water productivity in tropical river basins: Enhancing food security and livelihoods by managing water for fish. *Agricultural Water Management*, 80(1), 262–275.

Duong, L. T., Nahn, D. K., Rothius, A., Quang, P. M., Giau, T. Q., Chi, H. H., Thuy, L. T., Hoa, N. V., Sinh, L. X., (1998). Integrated rice–fish culture in the Mekong Delta of Vietnam: problems, constraints and opportunities for sustainable agriculture. In: Xuan, V.-T., Matsui, S. (Eds.), Development of Farming Systems in the Mekong Delta of Vietnam JIRCAS, CTU & CLRRI. Ho Chi Minh Publishing House, Ho Chi Minh, pp. 235–271.

Edwards, C. A. (1987). The concept of integrated systems in lower input/sustainable agriculture. *American Journal of Alternative Agriculture*, 2(04), 148–152.

Edwards, C. A. (1989). The importance of integration in sustainable agricultural systems. *Agriculture, Ecosystems and Environment*, 27, 25–35.

Edwards, P. (1993). Environmental issues in integrated agriculture-aquaculture and wastewater-fed culture systems. In: Pullin, R. S. V., Rosenthal, H. and Maclean, J. L., (Eds.) Environment and aquaculture in developing countries, pp. 139–170. ICLARM Conference Proceeding. International Center for Living Aquatic Resources Management, Manila.

Edwards, P. (1998). A systems approach for the promotion of integrated aquaculture. *Aquaculture Economics and Management*, 2 (1), 1–12.

Edwards, P. (2003). Philosophy, principles and concepts of integrated agri-aquaculture systems. In Gooley, G. J., and Gavine, F. M. (Eds.). *Integrated Agri-Aquaculture Systems: a resource handbook for Australian industry development*. Rural Industries Research and Development Corporation. http://www.aces.edu/dept/fisheries/education/documents/Integrated_Agri_Aquaculture_Systems.pdf.

Edwards, P. (2000). Aquaculture, poverty impacts and livelihoods. *Natural Resource Perspectives*, 56. http:// www.odi.org.uk/nrp/56.pdf.

Edwards, P., Pullin, R. S. V., Garner, J. A., (1988). Research and education for the development of integrated crop-livestock-fish farming system in the tropics. ICLARM studies and reviews 16. International Center for Living Aquatic Resources Management, Manila, pp. 53. http://www.worldfishcenter.org/libinfo/Pdf/Pub%20SR76%2016.pdf.

Fan, S., Pandya-Lorch, R. (Eds.). (2012). Reshaping agriculture for nutrition and health. International Food Policy Research Institute. Washington, DC. http://www.ifpri.org/sites/default/files/publications/oc69.pdf.

FAO (2001). Integrated agriculture-aquaculture: A primer. FAO Fisheries Technical Paper-407, Food and Agriculture Organization of the United Nations, Rome, Italy. http://www.fao.org/docrep/005/y1187e/y1187e00.htm#TopOfPage (accessed on 05.05.2014).

FAO (2005a). EASYPol. On-line resource materials for policy making. Analytical tools. Module 043. Commodity Chain Analysis. Constructing the Commodity Chain, Functional Analysis and Flow Charts. Source: www.fao.org/docs/up/easypol/330/cca_043EN.pdf (accessed on 01.11.2009).

FAO (2005b). EASYPol. On-line resource materials for policy making. Analytical tools. Module 044. Commodity Chain Analysis. Financial Analysis. www.fao.org/docs/up/easypol/331/CCA_044EN.pdf (accessed on 01.11. 2009).

FAO (2006). Strengthening national food control systems. Guidelines to assess capacity building needs, Rome, Italy.

FAO (2010a). The state of food insecurity in the world, addressing food insecurity in protracted crises. Food and Agriculture Organization of the United

Nations. Rome, Italy. http://www.fao.org/docrep/013/i1683e/i1683e.pdf (accessed on 01-03-2012).

FAO (2010b). Climate-smart agriculture: policies, practices and financing for food security, adaptation and mitigation. Food and Agriculture Organization of the United Nations (FAO), Rome, Italy. http://www.fao.org/docrep/013/i1881e/i1881e00.pdf.

FAO, IFAD and WFP. (2013). The state of food insecurity in the world 2013. The multiple dimensions of food security. Rome, FAO. http://www.fao.org/docrep/018/i3434e/i3434e.pdf.

Fasse, A., Grote, U., Winter, E. (2009). Value chain analysis methodologies in the context of environment and trade research (No. 429). Discussion papers// School of Economics and Management of the Hanover Leibniz University. https://www.econstor.eu/dspace/bitstream/10419/37104/1/609241915.pdf.

Feder, G., and Umali, D. (1993). The Adoption of Agricultural Innovations A Review. *Technological Forecasting and Social Change*, 43, 215–239.

Feder, G., Just, R. E., and Zilberman, D. (1985). The Adoption of Agricultural Innovations in Developing Countries: A Survey. *Economic Development and Cultural Change*, 33(2), 255–298.

Fernando, C. H. (1993). Rice field ecology and fish culture—an overview. *Hydrobiologia*, 259(2), 91–113.

Fernando, C. H., and Halwart, M. (2000). Possibilities for the integration of fish farming into irrigation systems. *Fisheries Management and Ecology*, 7(1–2), 45–54.

Finnveden, G., and Moberg, A. (2005). Environmental systems analysis tools–an overview. *Journal of Cleaner Production*, 13, 1165–1173.

Foster, A., and Rosenzweig, M. (1995). Learning by Doing and Learning from Others: Human Capital and Farm household Change in Agriculture. *Journal of Political Economy*, 103(6): 1176–1209.

Frei, M., and Becker, K. (2005a). Integrated rice–fish culture: Coupled production saves resources. *Natural Resources Forum*, 29 (2), 135–143.

Frei, M., and Becker, K. (2005b). Integrated rice–fish production and methane emission under greenhouse conditions. *Agriculture, ecosystems and environment*, 107(1), 51–56.

Frei, M., Razzak, M. A., Hossain, M. M., Oehme, M., Dewan, S., and Becker, K. (2007a). Performance of common carp, Cyprinus carpio L. and Nile tilapia, Oreochromis niloticus (L.) in integrated rice–fish culture in Bangladesh. *Aquaculture*, 262(2), 250–259.

Frei, M., Khan, M. A. M., Razzak, M. A., Hossain, M. M., Dewan, S., Becker, K. (2007b). Effects of a mixed culture of common carp, *Cyprinus carpio* L., and

Nile tilapia, *Oreochromis niloticus* (L.), on terrestrial arthropod population, benthic fauna, and weed biomass in rice Welds in Bangladesh. *Biological Control,* 41, 207–213.

Frei, M., Razzak, M. A., Hossain, M. M., Oehme, M., Dewan, S., and Becker, K. (2007c). Methane emissions and related physicochemical soil and water parameters in rice–fish systems in Bangladesh. *Agriculture, ecosystems and environment, 120*(2), 391–398.

Geer, T., Debipersaud, R., Ramlall, H., Settle, W., Chakalall, B. Joshi, R. Halwart, M. (2007). Introduction of aquaculture and other integrated production management practices to rice farmers in guyana and suriname. FAO Aquaculture Newsletter., No. 35, pp. 48–50. http://www.fao.org/docrep/013/a0595e/a0595e20.pdf.

Gereffi, G. (1994). The organization of buyer-driven global commodity chains: How US retailers shape overseas production networks. In G. Gereffi, and M. Korzeniewicz (Eds.), *Commodity chains and global capitalism* (pp. 95–123). Westport, CT: Praeger.

Gereffi, G. (1995). Global production systems and third world development. In B. Stallings (Ed.), *Global change, regional response: The new international context of development* (pp. 100–142). Cambridge, MA: Cambridge University Press.

Gereffi, G., Humphrey, J., Kaplinsky, R., Sturgeon, T. (2001). Globalization, value chains and development. *IDS Bulletin*, 32(3), 1–9.

Giap, D. H., Yi, Y., and Lin, C. K. (2005). Effects of different fertilization and feeding regimes on the production of integrated farming of rice and prawn Macrobrachium rosenbergii (De Man). *Aquaculture Research*, 36 (3), 292–299.

Gibbon, P. (2000). Back to the basics through delocalisation: the Mauritian garment industry at the end of the twentieth century. Working Paper 00.7. Copenhagen: Centre for Development Research.

Godfray, H. C. J., Beddington, J. R., Crute, I. R., Haddad, L., Lawrence, D., Muir, J. F., ... and Toulmin, C. (2010). Food security: the challenge of feeding 9 billion people. *Science, 327*(5967), 812–818.

Gomiero, T., M. Giampietro, S. G. F. Bukkens, C. Liewan, X. Jinze, (1999). Environmental and socioeconomic constraints to the development of freshwater fish aquaculture in China. *Critical Reviews in Plant Sciences*, 18 (3), 359–371.

Gooley, G. J., Gavine, F. M., (Eds.), (2003). Introduction to integrated agri-aquaculture systems in Australia. Integrated Agri-Aquaculture Systems A Resource Handbook for Australian Industry Development. RIRDC Project No. MFR-2A. Rural Industries Research and Development Corporation, Australia.

Graaff, J., Amsalu, A., Bodnár, F., Kessler, A., Posthumus, H., and Tenge, A. (2008). Factors influencing adoption and continued use of long-term soil and water conservation measures in five developing countries. *Applied Geography*, 28(4), 271–280.

Greene, W. H. (2003). Econometric Analysis, 5th Ed. Prentice-Hall, Upper Saddle River, NJ.

Griliches, Z. (1957). Hybrid corn: An exploration in the economics of technological change. *Econometrica*, 25(4): 501–522.

Guan, R. J. and Chen, Y. D. (1989). Reform and development of China's fisheries. FAO Fish Circulation, 822, 3.

Guilkey, D. K., and Murphy, J. L. (1993). Estimation and testing in the random effects probit model. *Journal of Econometrics*, 59(3), 301–317.

Gupta, M. V. and Mazid, M. A. (1993) Feasibility and potentials for integrated rice–fish systems in Bangladesh. Paper presented at the Twelve session of the FAO Regional Farm Management Committee for Asia and the Far East, Dhaka, Bangladesh.11–14 December1993, pp. 1–19.

Gupta, MV, Ahmed, M, Bimbao, M, Lightfoot, C. (1992). Socioeconomic impact and farmers' assessment of Nile tilapia (Oreochromis niloticus) culture in Bangladesh. ICLARM Technical Report No 35. 50 p.

Gupta, M. V., Rahman, M. A., Mazid, M. A., Sollows, J. D., (1996). Integrated agriculture-aquaculture: A way for food security for small farmers and better resource management and environment. In: Heidhues, F., Fadani, A. (Eds.), Food Security and Innovations—Successes and Lessons Learned. Peter Lang, Frankfurt. 165–175.

Gupta, M. V., Sollows, J. D., Mazid, M. A., Rahman, A., Hussain, M. G., Dey, M. M., (1998). Integrating Aquaculture with Rice Farming in Bangladesh: Feasibility and Economic Viability, Its Adoption and Impact (ICLARM Technical Report 55). ICLARM, Penang, Malaysia.

Guptill, A., Wilkins, J. R. (2002). Buying into the food system: trends in food retailing in the US and implications for local foods. *Agriculture and Human Values*, 19 (1), 39–51.

Gurung, T. B., Wagle, S. K. (2005). Revisiting underlying ecological principles of rice–fish integrated farming for environmental, economical and social benefits. *Our Nature*, 3(1), 1–12.

Gurung, T. B. (2012). Integrated aquaculture within agriculture irrigation for food security and adaptation to climate change. *Hydro Nepal: Journal of Water, Energy and Environment*, 11(1), 73–77.

Hagedorn, K. (2008). Particular requirements for institutional analysis in nature-related sectors. *European Review of Agricultural Economics*, 35(3), 357–384.

Halls, A. S., Hoggarth, D. D., Debnath, K., (1998). Impact of flood control schemes on river fish migrations and assemblages in Bangladesh. *Journal of Fish Biology*, 53, 358–380.

Halwart, M. (1995). Fish as Biocontrol Agents in Rice: The Potential of Common Carp (Cyprinus carpio) and Nile Tilapia (Oreochromis niloticus) (Tropical Agroecology 8), ed. Halwart M (Margraf Verlag, Weikersheim, Germany).

Halwart, M, Borlinghaus, M, Kaule, G. (1996). Activity pattern of fish in rice fields. *Aquaculture*, 145(1), 159–170.

Halwart, M., and M. V. Gupta (eds.). (2004). Culture of fish in rice fields. Rome, Italy and Penang, Malaysia: FAO and The WorldFish Center, 1–83. <http://www.worldfishcenter.org/Pubs/CultureOfFish/Culture-of-Fish.pdf> (accessed on 26.08.2013).

Halwart, M. (1998). Trends in rice–fish farming. *FAO Aquaculture Newsletter (FAO)*, 18, 3–11.

Halwart, M. (2006). Biodiversity and nutrition in rice-based aquatic ecosystems. *Journal of Food Composition and Analysis*, 19(6), 747–751.

Haque, M. M, Little, D. C, Barman, B. K, Wahab, M. A. (2010). The adoption process of rice field-based fish seed production in northwest Bangladesh: an understanding through quantitative and qualitative investigation. *Journal of Agricultural Education and Extension*, 16(2), 161–177.

Haroon, A. K. Y., Dewan, S., Karim, S. M. R. (1992). Rice–fish production systems in Bangladesh. In: dela Cruz, C. R., Lightfoot, C., Costa-Pierce, B. A., Carangal, V. R., Bimbao, M. A. P. (Eds.), *Rice–Fish Research and Development in Asia*, ICLARM Conference Proceedings 24. ICLARM, Manila, pp. 165–171.

Hayami, Y. and V. Ruttan. (1985). *Agricultural development: An international perspective*. Second edition. Baltimore, MD: John Hopkins University Press.

Hecht, J. E. (2007). National Environmental Accounting–A Practical Introduction. *International Review of Environmental and Resource Economics*, 1, 3–66.

Heckman, J. J., and Hotz, V. J. (1989). Choosing among alternative nonexperimental methods for estimating the impact of social programs: The case of manpower training. *Journal of the American statistical Association*, 84(408), 862–874.

Heckman, J. J., Ichimura, H., and Todd, P. E. (1997). Matching as an econometric evaluation estimator: Evidence from evaluating a job training programme. *The review of economic studies*, 64(4), 605–654.

Hendarsih, S., Suriapermana, S., Fagi, A. and Manwan, I. (1994). Potential of fish in rice–fish culture as a biological control agent of rice pests. In Role of Fish in Enhancing Ricefield Ecology and in Integrated Pest Management (Ed. C. R. dela Cruz), p. 32–33. ICLARM Conference Proceedings 43. Manila: ICLARM.

Heong, K. L., Cuc, N. T. T., Binh, N., Fujisaka, S., Bottrell, D. G. (1995). Reducing early-season insecticide applications through farmer's experiment in Vietnam. In: Denning, G. L., Xuan, V. T. (Eds.), Vietnam and IRRI: a Partnership in Rice

Research. International Rice Research Institute (IRRI), Manila, pp. 217–222. http://books.irri.org/9712200671_content.pdf.

Hilbrands, A., and Yzerman, C. (2004). On-farm fish culture. Agrodok 21. Digigrafi, Wageningen, the Netherlands. Second edition. 67 p. http://journeytoforever.org/farm_library/AD21.pdf.

Hofstede, A. E., and Ardiwinata, R. O. (1950). Compiling statistical data on fish culture in irrigated rice fields in West Java. *Landbouw, 22*, 469–492.

Holland, J. (2007). Tools for Institutional, Political, and Social Analysis of Policy Reform. A Source Book for Development Practitioners. The World Bank, Washington, D. C. http://siteresources.worldbank.org/EXTTOPPSISOU/Resources/1424002-1185304794278/TIPs_Sourcebook_English.pdf (accessed on 20.08.2013).

Horstkotte-Wesseler, G. (1999). *Socioeconomics of Rice-aquaculture and IPM in the Philippines: Synergies, Potential, and Problems.* ICLARM Tech. Rep. 57,225 p. http://books.irri.org/9718020047_content.pdf.

Hossain, M. (1988). The nature and impact of green revolution in Bangladesh. IFPRI Research Report 67, IFPRI, Washington, D. C. http://www.ifpri.org/sites/default/files/publications/rr67.pdf.

Hossain, M., (2010). Shallow tubewells, boro rice, and their impact on food security in Bangladesh. In: Spielman, D. J., Pandya-Lorch, R. (Eds.), *Proven Successes in Agricultural Development.* International Food Policy Research Institute, Washington, DC, pp. 243–270. http://environmentportal.in/files/Proven%20Successes.pdf#page=258.

Hossain, M., Bose, M. L., and Mustafi, B. A. (2006). Adoption and productivity impact of modern rice varieties in Bangladesh. *The Developing Economies, 44*(2), 149–166.

Hossain, M., Quasem, M. A., Akash, M. M. and Jabbar, M. A., (1990). Differential impact of modern rice technology: The Bangladesh case. Bangladesh Institute of Development Studies, Dhaka.

Hossain, S. T., Sugimoto, H., Ahmed, G. J. U., and Islam, M. R. (2005). Effect of integrated rice-duck farming on rice yield, farm productivity, and rice-provisioning ability of farmers. *Asian Journal of Agriculture and Development.* 2 (1), 79–86.

Huang, J., Hu, R., Rozelle, S., and Pray, C. (2005). Insect-resistant GM rice in farmers' fields: assessing productivity and health effects in China. *Science, 308*(5722), 688–690.

Huda, K. M., Atkins, P. J., Donoghue, D. N., and Cox, N. J. (2010). Small water bodies in Bangladesh. *Area, 42*(2), 217–227.

Humphrey, J. Schmitz, H. (2001). Governance in global value chains. *IDS Bulletin, 32*(3), 19–30.

Ichinose, K., Tochihara, M., Wada, T., Suguiura, N., and Yusa, Y. (2002). Influence of common carp on apple snail in a rice field evaluated by a predator-prey logistic model. *International Journal of Pest Management*, 48(2), 133–138.

IFAD. (2003). Indigenous Peoples and Sustainable Development. Roundtable Discussion Paper for the Twenty-Fifth Anniversary Session of IFAD's Governing Council. Rome, IFAD. http://www.ifad.org/gbdocs/gc/26/e/ip.pdf.

IFAD. (2011). Rural poverty report 2011. New realities, new challenges: new opportunities for tomorrow's generation. Rome, IFAD http://www.ifad.org/rpr2011/report/e/rpr2011.pdf.

IFPRI. (20110). Research and extension of rice–fish technology in Bangladesh: An expert opinion forum. A Cereal System Initiative for South Asia (CSISA) study on Developing and deploying new technologies to smallholders in South Asia: Key policies and issues. Forum report. International Food Policy Research Institute. Washington, DC.

Imbens, Guido W., and. Wooldridge. J. M. (2009). Recent developments in the econometrics of program evaluation. *Journal of Economic Literature*, 47(1), 5–86.

Jahan, K. Murshed-e, Crissman, C., and Antle, J. (2013). Economic and social impacts of Integrated Aquaculture-Agriculture technologies in Bangladesh. GIAR Research Program on Aquatic Agricultural Systems. WorldFish, Penang, Malaysia. Working Paper: AAS-2013-02. Penang, MALAYSIA. http://www.worldfishcenter.org/resource_centre/WF_3452.pdf.

Jahan, K. M., M. Beveridge, and A. C. Brooks. (2008). Impact of Long-term Training and Extension Support on Small-scale Carp Polyculture Farms of Bangladesh. *Journal of the World Aquaculture Society*, 39(4): 441–453.

Jahan, Murshed-E. K., and Pemsl, D. E. (2011). The impact of integrated aquaculture–agriculture on small-scale farm sustainability and farmers' livelihoods: Experience from Bangladesh. *Agricultural Systems*, 104(5), 392–402.

Jahan, Murshed-E. K., Crissman, C., and Antle, J. (2013). Economic and social impacts of Integrated Aquaculture-Agriculture technologies in Bangladesh. Penang, Malaysia, WorldFish, 14 pp. (Working Paper, AAS-2013-02).

Jamu, D. M., and Piedrahita, R. H. (2002). An organic matter and nitrogen dynamics model for the ecological analysis of integrated aquaculture/agriculture systems: II. Model evaluation and application. *Environmental Modelling and Software*, 17(6), 583–592.

Jianguo, W., and Dashu, N. (1995). A comparative study of the ability of fish to catch mosquito larva. In Rice–Fish Culture in China (Ed. K. T. MacKay), pp. 217–222. Ottawa: IDRC. http://www.idrc.ca/EN/Resources/Publications/openebooks/313-5/index.html#page_217.

Jorgenson, A., and Birkholz, R. (2010). Assessing the causes of anthropogenic methane emissions in comparative perspective, 1990–2005. *Ecological Economics, 69*(12), 2634–2643.

Kangmin, L. (1988). Rice–fish culture in China: A review. *Aquaculture, 71*(3), 173–186.

Kaplinsky, R. (2000). Globalisation and unequalisation: What can be learned from value chain analysis?. *Journal of development studies, 37*(2), 117–146.

Kaplinsky, R, Morris, M. (2001). *A handbook for value chain research* (Vol. 113). Ottawa: IDRC. http://www.prism.uct.ac.za/Papers/VchNov01.pdf.

Karim, M. (2006). The livelihood impacts of fishponds integrated within farming systems in Mymensingh District, Bangladesh. PhD Thesis. Institute of Aquaculture, University of Stirling, UK.

Kathage, J., and Qaim, M. (2012). Economic impacts and impact dynamics of Bt (Bacillus thuringiensis) cotton in India. *Proceedings of the National Academy of Sciences, 109*(29), 11652–11656.

Kawarazuka, N., and Béné, C. (2010). Linking small-scale fisheries and aquaculture to household nutritional security: an overview. *Food Security, 2*(4), 343–357.

Kay, R. D., William, M. E., Patricia, A. D. (2008). *Farm Management.* Sixth Edition. New York City: McGraw-Hill Companies.

Kerr, K., and Kolavalli, S. (1999). Impact of agricultural research on poverty alleviation: conceptual framework with illustrations from the literature. EPTD discussion paper no. 56. International Food Policy Research Institute. Washington, DC, USA. http://impact.cgiar.org/sites/default/files/KerrKolavalli1999.pdf (Accessed on 18.11.2011).

Khoo, K. H. and E. S. P. Tan. (1980). Review of rice–fish culture in Southeast Asia. In R. S. V. Pullin and Z. H. Shehadeh (eds.) Proceding of the ICLARM-SEARCA Conference on Integrated Agriculture-Aquaculture Farming Systems, 6–9 August 1979, Manila, Philippines, 258 p.

Kijima, Y., Otsuka, K., and Sserunkuuma, D. (2011). An inquiry into constraints on a green revolution in Sub-Saharan Africa: the case of NERICA rice in Uganda. *World Development, 39*(1), 77–86.

Kikulwe, E. M., Fischer, E., and Qaim, M. (2014). Mobile Money, Smallholder Farmers, and Household Welfare in Kenya. *PloS one, 9*(10), e109804.

Kim, S, Shin, E. (2002). A Longitudinal Analysis of Globalization and Regionalization in International Trade: A Social Network Approach. *Social Forces, 81*(2), 445–471.

Kipkemboi, J., Van Dam, A. A., Ikiara, M. M., and Denny, P. (2007). Integration of smallholder wetland aquaculture–agriculture systems (fingerponds) into riparian farming systems on the shores of Lake Victoria, Kenya: socio-economics and livelihoods. *The Geographical Journal, 173*(3), 257–272.

Kledal, P. R. (2006). The Danish organic vegetable chain, Report No 182 (Den Kgl. Veterinarog Lantbohojskole), Copenhagen.

Klemick, H., and Lichtenberg, E. (2008). Pesticide use and fish harvests in Vietnamese rice agroecosystems. *American Journal of Agricultural Economics, 90*(1), 1–14.

Kohls, R. L, Uhl, J. N. (2002). *Marketing of agricultural products* (No. Ed. 9). Prentice-Hall.

Koohanfkan, P, Furtado, J. (2004). Traditional rice–fish systems as globally indigenous agricultural heritage systems (GIAHS). Proceeding of the FAO Rice Conference. International Rice Commission Newsletter 53:66–74. http://www.fao.org/3/a-y5682e/y5682e0b.htm.

Koroma, S. (2007), Globalization, agriculture and the least developed countries. This issues paper was prepared for the Ministerial Conference "Making Globalization Work for the LDCS", Istanbul, Turkey, July 9–11, 2007 by Suffyan Koroma, Trade and Markets Division, Economic and Social Department, Food and Agriculture Organization of the United Nations (FAO), Rome, Italy. http://www.unohrlls.org/UserFiles/File/LDC%20Documents/Turkey/20June07-Agriculture-Final.pdf (accessed on 02-03-2012).

Kouser, S., and Qaim, M. (2011). Impact of Bt cotton on pesticide poisoning in smallholder agriculture: A panel data analysis. *Ecological Economics, 70*(11), 2105–2113.

Kremen, C., and Miles, A. (2012). Ecosystem services in biologically diversified versus conventional farming systems: benefits, externalities, and trade-offs. *Ecology and Society, 17*(4), 40.

Krishi Diary. (2011). Agricultural information service. Dhaka, Bangladesh.

Kunio, T. (2002). Views on food production: towards a new green revolution. Paper prepared for presentation at the 13th International Farm Management Congress, Wageningen, The Netherlands, July 7–12, 2002. http://ageconsearch.umn.edu/bitstream/6987/2/cp02ts01.pdf (accessed on 04.02.2012).

Kurttila, M., Pesonen M, Kangas J, Kajanus M. (2000). Utilizing the analytic hierarchy process (AHP) in SWOT analysis—a hybrid method and its application to a forest certification case. *Forest Policy and Economics,* 1(1), 41–52.

Langhu, W., (1995). Methods of rice–fish farming and their ecological efficiency, 91-96. In: MacKay, T. K. (Ed.), *Rice–Fish Culture in China.* International Development Research Centre (IDRC), Ottawa. http://www.idrc.ca/EN/Resources/Publications/openebooks/313-5/index.html#page_91.

Läpple, D. (2010). Adoption and abandonment of organic farming: an empirical investigation of the Irish drystock sector. *Journal of Agricultural Economics*, 61(3), 697–714.

Lee, D. K., and Lee, W. J. (2003). Biological control of mosquito larvae by using indigenous fish predator, Misgurnus mizolepis in Republic of Korea. In *Abstract book of the 52nd annual meeting of the American Society of Tropical Medicine and Hygiene* (Vol. 3). 3 to 7 December, Philadelphia.

Lee, D., Ruben, R., (2000). Putting the 'farmer first': returns to labor and sustainability in agro-ecological analysis. In: 24th International Conference of the International Agricultural Economics Association. Berlin. http://dyson.cornell.edu/research/researchpdf/wp/2000/Cornell_Dyson_wp0012.pdf.

Lenzen, M. (2001). Errors in Conventional and Input-Output-based Life-Cycle Inventories. *Journal of Industrial Ecology*, 4 (4), 127–149.

Leuven, E., Sianesi, B., (2003). PSMATCH2: Stata module to perform full Mahalanobis and propensity score matching, common support graphing, and covariate imbalance testing. https://ideas.repec.org/c/boc/bocode/s432001.html.

Li, K., (1986). A reviewof rice–fish culture in China.Network of Aquaculture Centers in Asia, Bangkok, Thailand and Freshwater Fisheries Research Centre, Chinese Academy of Fisheries Science Wuxi, China. http://www.fao.org/docrep/field/003/ac221e/ac221e00.htm.

Li, X., Wu, H. and Zhang, Y. (1995). Economic and ecological benefits of rice–fish culture. In *Rice–Fish Culture in China* (Ed K. T. MacKay), pp. 129–138. Ottawa: IDRC. http://www.idrc.ca/EN/Resources/Publications/openebooks/313-5/index.html#page_129.

Lightfoot, C, van Dam, A., Costa-Pierce, B. (1992). What's happening to rice yields in rice–fish systems?. In dela Cruz, C. R., Lightfoot, C., Costa-Pierce, B. A., Carangal, V. R. and Bimbao, M. P. (eds), *Rice–Fish Research and Development in Asia*. ICLARM Conference Proceedings 24, Manila, Philippines, pp. 177–183. <http://www.worldfishcenter.org/resource_centre/WF_102.pdf> (accessed on 26.08.2013).

Lightfoot, C. (1990). Integration of aquaculture and agriculture: a route to sustainable farming systems. *Naga, The ICLARM Quarterly*, 13(1), 9–12. http://www.worldfishcenter.org/Naga/na_2841.pdf.

Lightfoot, C., Bimbao, M. P., Dalsgaard, J. P. T., Pullin, R. S. V. (1993). Aquaculture and sustainability through integrated resources management. *Outlook on Agriculture*, 22 (3), 143–150.

Lipton, M. and R. Longhurst. (1989). New seeds and poor people. London: Unwin Hyman.

Little, D. C, Surintaraseree, P, Innes-Taylor, N. (1996). Fish culture in rainfed rice fields of northeast Thailand. *Aquaculture,* 140(4), 295–321.

Little, D., and Muir, J. (2003). Integrated agri-aquaculture systems-the Asian experience. In Gooley, G. J., and Gavine, F. M. (Eds.). *Integrated Agri-Aquaculture Systems: a resource handbook for Australian industry development.* Rural Industries Research and Development Corporation. http://www.aces.edu/dept/fisheries/education/documents/Integrated_Agri_Aquaculture_Systems.pdf.

Little, D. C., Muir, J., (1987). A guide to integrated warm water aquaculture. Institute of Aquaculture, University of Stirling, Stirling, UK.

Lu, J., and Li, X. (2006). Review of rice–fish-farming systems in China—one of the globally important ingenious agricultural heritage systems (GIAHS). *Aquaculture,* 260(1), 106–113.

Macfadyen, G., Nasr-Alla, A. M., Al–Kenawy, D, Fathi M, Hebicha, H, Diab A. M, Hussein, S M, Abou-Zeid, R. M, El-Naggar, G. (2012). Value-chain analysis–an assessment methodology to estimate Egyptian aquaculture sector performance. *Aquaculture,* 362–363, 18–27.

Maertens, M, Swinnen, J. F. (2010). Are African high-value horticulture supply chains bearers of gender inequality? In Workshop on Gaps, trends and current research in gender dimensions of agricultural and rural employment: differentiated pathways out of poverty (Vol. 31). Rome, Italy. http://www.fao.org/uploads/media/Gender%20issues.pdf.

Mahajan, V., and Peterson, R. A. (Eds.). (1985). *Models for innovation diffusion* (Vol. 48). California: Sage publication.

Mansfield, E. (1961). Technical change and the rate of imitation. *Econometrica,* 29, 741–766.

Marra, M., Pannell, D. J. and Ghadim, A. A. (2003), The economics of risk, uncertainty and learning in the adoption of new agricultural technologies: Where are we on the learning curve?. *Agricultural Systems,* 75(2–3), 215–234.

Mathias, J. A., Charles, A. T., Hu, B. T., (1998). Integrated fish farming. Proceedings of a Workshop on Integrated Fish Farming, 11–15 October, Wuxi, Jiangsu Province, China. CRC Press, Boca Raton, Florida, USA.

Matteson, P. C. (2000). Insect-pest management in tropical Asian irrigated rice fields. *Annual Review Entomology,* 5, 549–574.

McDonald, J. F., and Moffitt, R. A. (1980). The uses of Tobit analysis. *The review of economics and statistics,* 62(2): 318–321.

McFadden, D., (1974). Conditional logit analysis of quantitative choice behavior. In P. Zarembka (Ed.), *Frontiers in Econometrics.* New York: Academic Press, 105–142.

Meaden, G. J. and Kapetsky, J. M. (1991). Geographical information systems and remote sensing in inland fisheries and aquaculture: FAO Fisheries Technical Paper 318, Food and Agricultural Organisation, Rome, Italy.

Mellor, J. (1998), Agriculture on the Road to Industrialization. In Carl Eicher and John Staatz, eds., *International Agricultural Development*. Baltimore: Johns Hopkins University Press.

Mendola, M. (2007). Agricultural technology adoption and poverty reduction: A propensity-score matching analysis for rural Bangladesh. *Food policy*, *32*(3), 372–393.

Minten, B, Murshid, K. A. S, Reardon, T. (2011). The quiet revolution in agrifood value chains in Asia: The case of increasing quality in rice markets in Bangladesh (No. 1141). International Food Policy Research Institute (IFPRI). http://ideas.repec.org/p/fpr/ifprid/1141.html.

Minten, B, Murshid, K. A. S, Reardon, T. (2013). Food quality changes and implications: evidence from the rice value chain of Bangladesh. World Development, 42, 100–113.

Mishra, A., and Mohanty, R. K. (2004). Productivity enhancement through rice–fish farming using a two-stage rainwater conservation technique. *Agricultural Water Management*, 67, 119–131.

Mishra, A., and Salokhe, V. (2010). The effects of planting pattern and water regime on root morphology, physiology and grain yield of rice. *Journal of Agronomy and Crop Science*, 196(5), 368–378.

Mohanty, R. K., Verma, H. N., and Brahmanand, P. S. (2004). Performance evaluation of rice–fish integration system in rainfed medium and ecosystem. *Aquaculture*, 23, 125–135.

Montpellier Panel Report (2013). Sustainable intensification: a new paradigm for African agriculture, London. http://ag4impact.org/wp-content/uploads/2014/07/Montpellier-Panel-Report-2013-Sustainable-Intensification-A-New-Paradigm-for-African-Agriculture-1.pdf.

Moser, C. M., and Barrett, C. B. (2006). The complex dynamics of smallholder technology adoption: the case of SRI in Madagascar. *Agricultural Economics*, *35*(3), 373–388.

Moser, C. M., and C. B. Barrett. (2003). The disappointing adoption dynamics of a yield-increasing, low external-input technology: the case of SRI in Madagascar. *Agricultural Systems*, 76 (3), 1085–100.

Moulton, T. P. (1973). More rice and less fish: Some problems of the Green Revolution. *Australian natural history*, 6, 322–327.

Mukherjee, T. K. (1995). Integrated crop-livestock-fish production systems for maximizing productivity and economic efficiency of small holders' farms.

In: *The Management of Integrated Freshwater Agro-piscicultural Ecosystems in Tropical Areas* (ed. by J. J. Symoens and J. C. Micha), pp. 121–143. Seminar Proceedings, RAOS and CTA, Wageningen, Brussels, Belgium.

Mukhopadhyay, P. K., Das, D. N., Roy, B., (1992). On-farm research in deep-water rice–fish culture in West Bengal, India. In: dela Cruz, C. R., Lightfoot, C., Costa-pierce, B. A., Carangal, V. R., Bimbao, M. P. (Eds.), ICLARM Conference Proceedings of the Rice–Fish Research and Development in Asia, vol. 24, p. 457.

Mundlak, Y. (1978). On the pooling of time series and cross section data. *Econometrica*, 46 (1), 69–85.

Muriithi, B. W., and Matz, J. A. (2015). Welfare effects of vegetable commercialization: Evidence from smallholder producers in Kenya. *Food Policy*, 50, 80–91.

Nabi, R. (2008). Constraints to the adoption of rice–fish farming by smallholders in Bangladesh: a farming systems analysis. *Aquaculture Economics and Management*, 12(2), 145–153.

Nahar, A. (2010). Impacts of different rice-fish-prawn culture systems on yield of rice, fish and prawn and limnological conditions. *Journal of the Bangladesh Agricultural University*, 8(1), 179–185.

Naylor, R. L., Goldburg, R. J., Primavera, J. H., Kautsky, N., Beveridge, M. C. M., Clay, J., Folke, C., Lubchenco, J., Mooney, H., Troell, M., (2000). Effect of aquaculture on world fish supplies. *Nature*, 405, 1017–1024.

Negatu, W., and Parikh, A. (1999). The impact of perception and other factors on the adoption of agricultural technology in the Moret and Jiru *Woreda* (district) of Ethiopia. *Agricultural Economics*, 21(2), 205–216.

Neill, S. P. and Lee, D. R. (2001). Explaining the adoption and disadoption of sustainable agriculture: The case of cover crops in northern Honduras. *Economic Development and Cultural Change*, 49 (4), 793–840.

Neng, W Guohou, L Yulin, L., and Gemei, Z (1995). The role of fish in controlling mosquitoes in rice fields. In *Rice–Fish Culture in China* (Ed. K. T. MacKay), pp. 213–216. Ottawa: IDRC. http://www.idrc.ca/EN/Resources/Publications/openebooks/313-5/index.html#page_213.

Nhan, D. K, Phong, L. T, Verdegem, M. J, Duong, L. T, Bosma, R. H, Little, D. C. (2007). Integrated freshwater aquaculture, crop and livestock production in the Mekong delta, Vietnam: determinants and the role of the pond. *Agricultural systems*, 94(2), 445–458.

Nie, D., Chen, Y., Wang, J., (1992). Mutualism of rice and fish in rice fields. In: dela Cruz, C. R., Lightfoot, C., Costa-Pierce, B. A., Carangal, V. R., Bimbao, M. P. (Eds.), ICLARM Conference Proceedings of the Rice–Fish Research and Development in Asia, p. 457. http://pdf.usaid.gov/pdf_docs/PNABQ682.pdf.

Nix, J. (2000). *Farm Management Pocketbook*, 31st Edition. Wye College, London, 244 pp.

Noltze, M., Schwarze, S., and Qaim, M. (2012). Understanding the adoption of system technologies in smallholder agriculture: The system of rice intensification (SRI) in Timor Leste. *Agricultural systems, 108*, 64–73.

Noltze, M., Schwarze, S., and Qaim, M. (2011). The System of Rice Intensification (SRI) in Timor Leste. Pacific News, 35, 4–9.

Noorhosseini, S. A., and Radjabi, R. (2010). Decline application of insecticide and herbicides in integrated rice–fish farming: The case study in north of Iran. *American Eurasian Journal of Agricultural and Environmental Science, 8*(3), 334–338.

Noorhosseini-Niyaki, S. A., and Allahyari, M. S. (2012). Logistic Regression Analysis on Factors Affecting Adoption of Rice–Fish Farming in North Iran. *Rice Science, 19*(2), 153–160.

Ofori, J., Abban, E. K., Otoo, E., and Wakatsuki, T. (2005). Rice–fish culture: an option for smallholder Sawah rice farmers of the West African lowlands. *Ecological Engineering, 24*(3), 233–239.

Omidi-Najafabadi, M., and Masjedi, S. H. K. (2011). Extension challenges and requirements of integrated rice–fish farming in Gilan province, Iran. *International Journal of Agricultural Science and Research, 2*(1), 1–8.

Ondersteijn, C. J, Wijnands, J. H, Huirne, R. B. (Eds.) (2006). Quantifying the agri-food supply chain (Vol. 15). Wageningen, Springer-Verlag.

Pacini, C., Giesen, G., Wossink, A., Omodei-Zorini, L., Huirne, R. (2004). The EU's Agenda 2000 reform and the sustainability of organic farming in Tuscany: ecological-economic modelling at field and farm level. *Agricultural Systems, 80*(2), 171–197.

Pant, J., Barman, B. K., Murshed-E-Jahan, K., Belton, B., Beveridge, M. (2014). Can aquaculture benefit the extreme poor? A case study of landless and socially marginalized Adivasi (ethnic) communities in Bangladesh. *Aquaculture, 418*, 1–10.

Pant, J., Demaine, H., and Edwards, P. (2005). Bio-resource flow in integrated agriculture–aquaculture systems in a tropical monsoonal climate: a case study in Northeast Thailand. *Agricultural systems, 83*(2), 203–219.

Papke, L. E., and Wooldridge, J. M. (2008). Panel data methods for fractional response variables with an application to test pass rates. *Journal of Econometrics, 145*, 121–133.

Patra, B. C., and Sinhababu, D. P. (1995). Weeds in rainfed lowland rice–fish system. *Oryza, 32*, 121–123.

Pender, J. (2007). Agricultural technology choices for poor farmers in less-favored areas of South and East Asia. IFPRI Discussion Paper 00709. Environment and Production Technology Division. International Food Policy Research Institute. Washington, DC. http://www.ifpri.org/sites/default/files/publications/ifpridp00709.pdf (Accessed on 28.10.2011).

Perales, F. F., (2013). MUNDLAK: Stata module to estimate random-effects regressions adding group-means of independent variables to the model. https://ideas.repec.org/c/boc/bocode/s457601.html (accessed on 01.12.2014).

Phong, L. T., van Dam, A. A., Udo, H. M. J., Van Mensvoort, M. E. F., Tri, L. Q., Steenstra, F. A., and Van der Zijpp, A. J. (2010). An agro-ecological evaluation of aquaculture integration into farming systems of the Mekong Delta. *Agriculture, ecosystems and environment*, 138(3), 232–241.

Pimentel, D., and Pimentel, M. (1990). Comment: Adverse environmental consequences of the Green Revolution. *Population and Development Review*, 16(Supplement): 329–332.

Pingali, P. L. (2012). Green Revolution: Impacts, limits, and the path ahead. *Proceedings of the National Academy of Sciences*, 109(31), 12302–12308.

Pingali, P. L., and Rosegrant, M. W. (1994). *Confronting the environmental consequences of the Green Revolution in Asia* (No. 2). EPTD discussion paper No. 2 International Food Policy Research Institute (IFPRI), Washington, DC, USA. http://www.ifpri.org/sites/default/files/publications/eptdp02.pdf.

Plant, R. (2002). Indigenous peoples/ethnic minorities and poverty reduction. Regional report. Environment and Social Safeguard Division. Regional and Sustainable Development Department, Asian Development Bank, Manila, Philippines.

Prein, M. (2002). Integration of aquaculture into crop–animal systems in Asia. *Agricultural systems*, 71(1), 127–146.

Prein, M. (1998). Rice–Fish Culture. International Development Research Centre (IDRC), Ottawa, Canada. http://www.yale.edu/macmillan/pier/resources/lessons/apdxii.pdf.

Prein, M., and Ahmed, M. (2000). Integration of aquaculture into smallholder farming systems for improved food security and household nutrition. *Food and Nutrition Bulletin*, 21(4), 466–471.

Prein, M., R. Oficial, M. Bimao and T. Lopez, (1998). Aquaculture for Diversification of Small Farms within Forest Buffer Zone management: an example from the Uplands of Quirino Province, Philippines. In: P. Edwards, et al. (Eds.), *Rural Aquaculture*. CAB International, Chiang Mai, Thailand, pp. 97–110.

Primavera, J. H. (2006). Overcoming the impacts of aquaculture on the coastal zone. *Ocean and Coastal Management*, 49, 531–545.

PRSP. (2008). Moving ahead: National strategy for accelerated poverty reduction II (FY 2009–11). General Economics Division, Planning Commission, Government of the People's Republic of Bangladesh, Dhaka, Bangladesh. http://www.lcgbangladesh.org/aidgov/WorkShop/2nd%2020PRSP%2020Final%20 20%28October-2008%29.pdf.

Pullin, R. S. V. (1998). Aquaculture, integrated resources management and the environment. In: Integrated Fish Farming (eds. J. A. Mathias, A. T. Charles and H. Baotong), Proceedings of a Workshop on Integrated Fish Farming, 11–15 October 1994, Wuxi, Jiangsu Province, China. CRC Press, Boca Raton, Florida, USA, pp.19–43.

Pullin, R. S. V., Shehadeh, Z., (1980). Integrated agriculture-aquaculture farming systems. ICLARM Conference Proceedings 4. ICLARM, Manila.

Purba, S., (1998). The economics of rice–fish production systems in North Sumatra, Indonesia: An empirical and model analysis. University of Göttingen, Diss, 188 pp.

Rahm, M. R., and Huffman, W. E. (1984). The adoption of reduced tillage: the role of human capital and other variables. *American journal of agricultural economics*, 66(4), 405–413.

Rahman, S. (2013). Formalin–never ending woe. New Age. Bangladesh Daily English Newspaper. http://www.newagebd.com/supliment.php?sid=183.

Rahman, S. (2003a). Environmental impacts of modern agricultural technology diffusion in Bangladesh: an analysis of farmers' perceptions and their determinants. *Journal of environmental management*, 68(2), 183–191.

Rahman, S. (2003b). Farm-level pesticide use in Bangladesh: determinants and awareness. *Agriculture, ecosystems and environment*, 95(1), 241–252.

Rahman, S. (2005). Environmental impacts of technological change in Bangladesh agriculture: farmers' perceptions, determinants, and effects on resource allocation decisions. *Agricultural economics*, 33(1), 107–116.

Rahman, S., Barmon, B. K., and Ahmed, N. (2011). Diversification economies and efficiencies in a 'blue-green revolution' combination: a case study of prawn-carp-rice farming in the 'gher' system in Bangladesh. *Aquaculture International*, 19, 665–682.

Rahman, S., and Thapa, G. B. (1999). Environmental impacts of technological change in Bangladesh agriculture: farmers' perceptions and empirical evidence. *Outlook on Agriculture*, 28(4), 233–238.

Reardon, T, Chen, K, Minten, B, Adriano, L. (2012). The quiet revolution in staple food value chains.: enter the Dragon, the Elephant, and the Tiger. Asian Development Bank (ADB)/IFPRI, Manila/Washington, DC. http://www.ifpri.org/sites/default/files/publications/quiet-revolution-staple-food-value-chains.pdf.

Rebitzer, G, Ekvall, T, Frischknecht, R, Hunkeler, D, Norris, G, Rydberg, T, Schmid, W. P, Suh S, Weidema, B. P, Pennington, D. W. (2004). Life cycle assessment: Part 1: Framework, goal and scope definition, inventory analysis, and applications. *Environment international*, 30(5), 701–720.

Redclift, M. (1989). The environmental consequences of Latin America's agricultural development: some thoughts on the Brundtland Commission report. *World Development*, 17(3), 365–377.

Riisgaard, L., Bolwig, S., Ponte, S., Du Toit, A, Halberg, N., Matose, F. (2010). Integrating poverty and environmental concerns into value-chain analysis: a strategic framework and practical guide. *Development Policy Review*, 28(2), 195–216.

Rola, A. C., and P. L. Pingali. (1993). Pesticides, rice productivity and farmer's health. Manila, Philippines: IRRI.

Röling, N., (2009). Pathways for impact: scientists' different perspectives on agricultural innovation. *International Journal of Agricultural Sustainability*, 7, 83–94.

Roos, N., Islam, M. M., Thilsted, S. H. (2003). Small indigenous fish species in Bangladesh: contribution to vitamin A, calcium and iron intakes. *The Journal of nutrition*, 133(11), 4021S–4026S.

Roos, N., Wahab, M. A., Chamnan, C., Thilsted, S. H. (2007a). The role of fish in food-based strategies to combat vitamin A and mineral deficiencies in developing countries. *The Journal of Nutrition*, 137(4), 1106–1109.

Roos, N., Wahab, M., Hossain, M. A. R., Thilsted, S. H. (2007b). Linking human nutrition and fisheries: incorporating micronutrient-dense, small indigenous fish species in carp polyculture production in Bangladesh. *Food and Nutrition Bulletin*, 28(Supplement 2), 280S–293S.

Rosenbaum, P. R., and Rubin, D. B. (1983). The central role of the propensity score in observational studies for causal effects. *Biometrika*, 70(1), 41–55.

Rosenbaum, P. R., and Rubin, D. B. (1985). Constructing a control group using multivariate matched sampling methods that incorporate the propensity score. *The American Statistician*, 39(1), 33–38.

Roth, S., Hyde, J. (2002). Partial budgeting for agricultural businesses. Pennsylvania state university, unpublished manual. http://pubs.cas.psu.edu/freepubs/pdfs/ua366.pdf (accessed on 10.02.2014).

Rothuis, A. J., Nhan, D. K., Richter, C. J. J, Ollevier, F. (1998b). Rice with fish culture in the semi-deep waters of the Mekong Delta, Vietnam: interaction of rice culture and fish husbandry management on fish production. *Aquaculture Research*, 29(1), 59–66.

Rothuis, A. J., Nhan, D. K., Richter, C. J., Ollevier, F. (1998a). Rice with fish culture in the semi-deep waters of the Mekong Delta, Vietnam: a socio-economical survey. *Aquaculture research*, 29(1), 47–57.

Rothuis, A. J, Vromant, N., Xuan, V. T., Richter, C. J. J, Ollevier F. (1999). The effect of rice seeding rate on rice and fish production, and weed abundance in direct-seeded rice–fish culture. *Aquaculture*, 172(3), 255–274.

Roy, R. D. (2012). Country Technical Notes on Indigenous Peoples' Issues: People's republic of Bangladesh. International Fund for Agricultural Development (IFAD), Rome, Italy. http://www.ifad.org/english/indigenous/pub/documents/tnotes/bangladesh.pdf.

Ruddle, K. and Zhong, G. F. (1988). *Integrated agriculture-aquaculture in South China. The dike-pond system of the Zhujiang Delta.* Cambridge: Cambridge University Press.

Ruddle, K. (1982). Traditional integrated farming systems and rural development: the example of rice-field fisheries in southeast Asia. *Agricultural Administration*, 10, 1–11.

Ruel, M. T., Alderman, H. (2013). Nutrition-sensitive interventions and programmes: how can they help to accelerate progress in improving maternal and child nutrition?. *The Lancet*, 382(9891), 536–551.

Saikia, S. K., and Das, D. N. (2008). Rice–fish culture and its potential in rural development: A lesson from Apatani farmers, Arunachal Pradesh, India. *Journal of Agriculture and Rural Development*, 6(1), 125–131.

Salter, A. J., and Martin, B. R. (2001). The economic benefits of publicly funded basic research: a critical review. *Research policy*, 30(3), 509–532.

Samsuzzaman, S., (2002). Integrated Homestead Farming. North Bangle Institute, RDRS, Rangpur, Bangladesh, Bangladesh.

Sarder, R. (2007). FAO Fisheries technical paper, No. 501. In: Bondad-reantaso, M. G. (Ed.), Freshwater fish seed resources in Bangladesh: assessment of freshwater fish seed resources for sustainable aquaculture. FAO, Italy, Rome, pp. 105–128.

Sarker, M. A. R., Alam, K., and Gow, J. (2012). Exploring the relationship between climate change and rice yield in Bangladesh: An analysis of time series data. *Agricultural Systems*, 112, 11–16.

Satari, G., (1962). Wet rice cultivation with fish culture: a study of some agronomical aspects, Ph.D. Thesis. Bogor Agricultural University, Bogor, Indonesia, 126 p.

Schoenly, K., Cohen, J. E., Heong, K. L., Litsinger, J. A., Aquino, G. B., Barrion, A. T., and Arida, G. (1996). Food web dynamics of irrigated rice fields at five elevations in Luzon, Philippines. *Bulletin of Entomological Research*, 86(04), 451–466.

Shang, Y. C. (1986). Research on aquaculture economics: a review. *Aquacultural Engineering*, 5(2), 103–108.

Shankar, V., Halls, A., Barr, J., (2004). Fish versus fish revisited: on the integrated management of floodplain resources in Bangladesh. *Natural Resources Forum*, 28 (2), 91–101.

Sharma, B. K. and Olah, J. (1986). Integrated fish-pig farming with agriculture in India and Hungary. *Aquaculture*, 54, 135–139.

Shiva, V. (1991). *The violence of Green Revolution: third world agriculture, ecology and politics*. Zed Books. London.

Singh, R. B. (2000). Environmental consequences of agricultural development: a case study from the Green Revolution state of Haryana, India. *Agriculture, ecosystems and environment*, 82(1), 97–103.

Sinhababu, D. P., Ghosh, B. C., Panda, M. M., Reddy, B. B. (1983). Effect of fish on growth and yield under rice–fish culture. *Oryza*, 20: 144–150.

Smith, J. A, and Todd, P. E. (2005). Does matching overcome LaLonde's critique of nonexperimental estimators?. *Journal of econometrics*, 125(1), 305–353.

Snijders, TAB. (2005). Fixed and random effects. In: Everitt BS, Howell DC (eds.) *Encyclopedia of Statistics in Behavioral Science*, 2, 664–665 (Wiley, Chichester).

Sollow, J. (2000). Rice–Fish Culture: Where and could it work?. Mekong fish: Catch and Culture. Mekong Fisheries Network Newsletter., 5 (3). http://www.mekonginfo.org/assets/midocs/0001641-biota-rice-fish-culture-where-and-could-it-work.pdf.

Spielman, D., and Pandya-Lorch, R. (2009). Millions fed: Proven successes in agricultural development. International Food Policy Research Institute (IFPRI). Washington, DC. http://www.ifpri.org/sites/default/files/publications/oc64.pdf (accessed on 15.02.2012).

Sturgeon, T. J. (2001). How do we define value chains and production networks?. *IDS bulletin*, 32(3), 9–18.

Subramanian, A., and Qaim, M. (2009). Village-wide effects of agricultural biotechnology: The case of Bt cotton in India. *World Development*, 37(1), 256–267.

Sugunan, V. V., Prein, M., Dey, M. M., (2006). Integrating agriculture, fisheries and ecosystem conservation: Win-win solutions. *International Journal of Ecology and Environmental Science*, 32, 3–14.

Surridge, C. (2004). Rice cultivation: Feast or famine?. *Nature*, 428, 360–361.

Syamsiah, I., Suriapermana, S., Fagi, A. M., (1992). Research on rice–fish culture: past experiences and future research program, p. 287–293. In: dela Cruz, C. R., Lightfoot, C., Costa-Pierce, B. A., Carangal, V. R., Carangal, V. R., Bimbao, M. P. (Eds.), ICLARM Conference Proceedings of the Rice–Fish Research and Development in Asia, vol. 24, p. 457.

Symones, J. J. and Micha, J. C. (1995). The management of integrated freshwater agro-pischiculture eco-systems in tropical area. In: Anonymous Proceedings on the international seminar held in Brussels, Belgium May 14–19 1994, The technical centre of Agricultural and Rural Co-operation (CTA), Wageningen, the Netherlands and the Belgium Royal Academy of Overseas Sciences (ARSOM), Brussels, Belgium.

Talpaz, H. and Tsur., Y. (1982). Optimizing aquaculture management of a single species fish population. *Agricultural Systems*, 9, 127–142.

Tatlıdil, F. F., Boz, İ., and Tatlidil, H. (2009). Farmers' perception of sustainable agriculture and its determinants: a case study in Kahramanmaras province of Turkey. *Environment, development and sustainability*, 11(6), 1091–1106.

Tauli-Corpuz, V. and Malanes, M. B. (2010). Overview: Assessing the First Decade of the World's Indigenous People (1995–2004): Volume II – The South Asia Experience. Tebtebba Foundation. Philippines. http://tebtebba.org/index.php/all-resources/category/8-books?download=824:assessing-the-first-decade-of-the-worlds-indigenous-people-1995-2004-volume-2.

Taylor, S. R., Pakdee, B. and Klampratum, D. (1988). Border method and fish culture: synergetic effects on the yield of rice grain. In The Second International Symposium on Tilapia in Aquaculture (Eds R. S. V. Pullin, T. Bhukaswan, K. Tonguthai and J. L. MacLean), pp. 91–98. ICLARM conference proceedings 15. Department of Fisheries, Bangkok, Manila: ICLARM.

Thilsted, S. H., Roos N., Hassan, N. (1997). The role of small indigenous fish species in food and nutrition security in Bangladesh. *Naga, The ICLARM Quarterly*, 20(3–4), 82–84.

Thirtle. C, Lin. L and J. Piesse (2003). The impact of research-led agricultural productivity growth on poverty reduction in Africa, Asia and Latin America. *World Development*, 31 (12), 1959–1975.

Thongpan, N., Singreuang, M., Thaila, C., Kaeowsawat, S., Sollows, J. D., (1992). On-farm rice–fish farming research in Ubonratchathani Province, Northeast Thailand. In: dela Cruz, C. R., Lightfoot, C., Costa-Pierce, B. A., Carangal, V. R., Bimbao, M. P. (Eds.), ICLARM Conference Proceedings of the Rice–Fish Research and Development in Asia, vol. 24, p. 457.

Tipraqsa, P., Craswell, E. T., Noble, A. D., and Schmidt-Vogt, D. (2007). Resource integration for multiple benefits: multifunctionality of integrated farming systems in Northeast Thailand. *Agricultural Systems*, 94(3), 694–703.

Tobin, J. (1958). Estimation of relationships for limited dependent variables. *Econometrica*, 26(1) 24–36.

Torres, J. N., Macabale, N. A., Mercado, J. R., (1992). On-farm rice–fish farming systems research in Guimba, Nueva Ecija, Philippines. Case Studies on Simultaneous and Rotational Rice–Fish Culture, pp. 208–223.

Tran, N., Crissman, C., Chijere, A., Hong, M. C., Teoh, S. J., and Valdivia, R. O. (2013). Ex-ante assessment of integrated aquaculture-agriculture adoption and impact in Southern Malawi. GIAR Research Program on Aquatic Agricultural Systems. Penang, Malaysia. Working Paper: AAS-2013-03. http://pubs.iclarm.net/resource_centre/WF_3453.pdf (accessed on 10.01.2015).

Trifković, N. (2014). Governance strategies and welfare effects: vertical integration and contracts in the catfish sector in Vietnam. *Journal of Development Studies*, 50 (7), 949–961.

Troell, M., Naylor, R. L., Metian, M., Beveridge, M., Tyedmers, P. H., Folke, C., Arrow, K. J., Barrett, S., Crépin, A. S., Ehrlich, P. R., Gren, A., Kautsky, N., Levin, S. A., Nyborg, K., Österblom, H., Polasky, S., Scheffer, M., Walker, B. H., Xepapadeas, T., de Zeeuw, A. (2014). Does aquaculture add resilience to the global food system?. *Proceedings of the National Academy of Sciences*, 111(37), 13257–13263.

Tsuruta, T., Yamaguchi, M., Abe, S. I., and Iguchi, K. I. (2011). Effect of fish in rice-fish culture on the rice yield. *Fisheries Science*, 77(1), 95–106.

Uddin, R, Wahid., M. I, Jasmeen, T., Huda, N. H., Sutradhar, K. B. (2011). Detection of formalin in fish samples collected from Dhaka city, Bangladesh. *Stamford Journal of Pharmaceutical Sciences*, 4(1), 49–52.

Van Dam, A. A. (1990). Multiple regression analysis of accumulated data from aquaculture experiments: a rice–fish culture example. *Aquaculture Research*, 21(1), 1–15.

Veliu, A., Gessese, N., Ragasa, C., Okali C. (2009). Gender analysis of aquaculture value chain in Northeast Vietnam and Nigeria. World Bank Agriculture and Rural Development Discussion Paper, 44. <http://siteresources.worldbank.org/INTARD/Resources/Gender_Aquaculture_web.pdf> (accessed on 26.08.2013).

Von Braun, J. (2007). The world food situation: new driving forces and required actions. Food policy report. International Food Policy Research Institute. Washington, D. C. http://www.ifpri.org/sites/default/files/pubs/pubs/fpr/pr18.pdf (accessed on 16.10.2011).

Von Braun, J. (1988). Effects of technological change in agriculture on food consumption and nutrition: rice in a West African setting. *World Development*, 16(9), 1083–1098.

Vromant, N., Duong, L. T., and Ollevier, F. (2002). Effect of fish on the yield and yield components of rice in integrated concurrent rice–fish systems. *The Journal of Agricultural Science*, 138(01), 63–71.

Vromant, N., Rothuis, A. J., Cuc, N. T. T., and Ollevier, F. (1998). The effect of fish on the abundance of the rice caseworm Nymphula depunctalis (Guenée)

(Lepidoptera: Pyralidae) in direct seeded, concurrent rice–fish fields. *Biocontrol Science and Technology*, 8(4), 539–546.

Wahab, M. A., Kunda, M., Azim, M. E., Dewan, S., and Thilsted, S. H. (2008). Evaluation of freshwater prawn-small fish culture concurrently with rice in Bangladesh. *Aquaculture Research*, 39(14), 1524–1532.

Walters, D., Lancaster, G. (2000). Implementing value strategy through the value chain. *Management Decision*, 38(3), 160–178.

Way, M. J., and Heong, K. L. (1994). The role of biodiversity in the dynamics and management of insect pests of tropical irrigated rice—a review. *Bulletin of Entomological Research*, 84(04), 567–587.

WDI (2014). World development indicators. The World Bank. http://elibrary.worldbank.org/doi/pdf/10.1596/978-1-4648-0163-1.

Weidner, M. (2011). Semiparametric estimation of nonlinear panel data models with generalized random effects. Working Paper, University College London. http://www.ifs.org.uk/conferences/weidnerrandomeffects.pdf.

William, J. G., Hella J. P., Mwatawala, M. W. (2012). Ex-ante economic impact assessment of green manure technology in maize production systems in Tanzania. *Research on Humanities and Social Sciences*, 2(9), 47–58.

Williamson, O. E. (1985). The Economic Institutions of Capitalism: Firms, Markets, Relational Contracting. London: Collier Maximilian Publisher.

Wilson, C., and Tisdell, C. (2001). Why farmers continue to use pesticides despite environmental, health and sustainability costs. *Ecological economics*, 39(3), 449–462.

Wood, A. (2001). Value chains: an economist's perspective. *IDS Bulletin*, 32(3), 41–45.

Wooldridge, J. M. (2010a). Correlated random effects models with unbalanced panels. Department of Economics. Michigan State University, USA.

Wooldridge, J. M. (2010b). *Econometric Analysis of Cross Section and Panel Data, 2nd Edition.* The MIT press, Cambridge, MA.

World Bank. (2012). Bangladesh: Annual economic update. Poverty Reduction and Economic Management, South Asia Region. The World Bank.

Wu, J. J., Babcock, B. A., (1998). The Choice of Tillage, Rotation, and Soil Testing Practices: Economic and Environmental Implications. *American Journal of Agricultural Economics,* 80, 494–511.

Wu, L. (1995). Methods of rice–fish culture and their ecological efficiency. In Rice–Fish Culture in China (Ed. K. T. MacKay), pp. 91–96. Ottawa: IDRC. http://www.idrc.ca/EN/Resources/Publications/openebooks/313-5/index.html#page_91.

Xie, J., Hu, L., Tang, J., Wu, X., Li, N., Yuan, Y., Yang, H., Zhang, J., Shiming Luo, S., and Chen, X. (2011a). Ecological mechanisms underlying the sustainability of the agricultural heritage rice–fish coculture system. *Proceedings of the National Academy of Sciences*, *108*(50), E1381–E1387.

Xie, J., Wu, X., Tang, J. J., Zhang, J. E., Luo, S. M., and Chen, X. (2011b). Conservation of traditional rice varieties in a globally important agricultural heritage system (GIAHS): Rice–fish co-culture. *Agricultural Sciences in China*, *10*(5), 754-761.

Xu, Y. and Guo, Y. (1992). Rice–fish farming systems research in China. In *Rice–fish research and development in Asia* (Eds C. R. dela Cruz, C. Lightfoot, B. A. Costa- Pierce, V. R. Carangal & M. P. Bimbao), pp. 315–323. ICLARM Conference Proceedings 24. Manila: ICLARM.

Yousuf, H., Dewan, A. K. S., Karim, S. M. R., (1992). Rice–fish production system in Bangladesh. In: dela Cruz, C. R., Lightfoot, C., Costa-Pierce, B. A., Carangal, V. R., Bimbao, M. P. (Eds.), ICLARM Conference Proceedings of the Rice–Fish Research and Development in Asia, vol. 24.

Yunus, M., Hardjamulia, A., Syamsiah, I. & Suriapermana, S. (1992). Evaluation of rice–fish production systems in Indonesia. In *Rice–Fish Research and Development in Asia* (Eds C. R. dela Cruz, C. Lightfoot, B. A. Costa-Pierce, V. R. Carangal and M. P. Bimbao), pp. 131–137. ICLARM Conference Proceedings 24. Manila: ICLARM.

Zepeda, L. (1994). Simultaneity of technology adoption and productivity. *Journal of Agricultural and Resource Economics*, 19 (1) 46-57.

Appendix

Table A.1.1: Attrition bias test results

Variables	Coef.	Std. Err.	z	P>z
Gender	0.04	0.31	0.13	0.90
Age	0.02	0.01	2.82	0.01
Marital Status	−0.05	0.25	−0.19	0.85
Education	0.00	0.02	−0.31	0.76
Main Occupation	0.05	0.16	0.31	0.75
Total family size	−0.03	0.04	−0.71	0.48
Farm size	0.00	0.00	1.61	0.11
Access to CBO	0.27	0.20	1.32	0.19
Access to extension	0.11	0.27	0.41	0.68
Access to irrigation	−0.17	0.16	−1.07	0.29
Irrigated area	0.00	0.00	0.33	0.75
Access to credit	−0.08	0.18	−0.45	0.65
Access to market information	−0.02	0.13	−0.14	0.89
Constant	0.25	0.51	0.48	0.63
Log likelihood	=	−243.68914		
LR chi2(13)	=	22.29		
Prob > chi2	=	0.05		
Pseudo R2	=	0.04		
Number of obs	=	656.00		

Table A.3.1: Mean and standard deviations (in parentheses) of independent variables by IAA Value chain participation category

Variables	Type	Definition and measurement	Total	Participator	Dis-participator	Non-participator	†	Diff ††	†††
Education	Continuous	Schooling of household head in Years	3.22 (3.85)	3.81 (4.02)	2.54 (3.48)	3.33 (3.98)	0.48	-0.79*	1.27***
Occupation	Dummy	=1 if the main occupation of the household head is agriculture	0.39 (0.49)	0.44 (0.50)	0.32 (0.47)	0.44 (0.50)	-0.01	-0.12**	0.12***
Marital status	Dummy	=1 if the household head is married	0.92 (0.28)	0.94 (0.23)	0.91 (0.28)	0.87 (0.34)	0.08*	0.05	0.03
Gender	Dummy	=1 if male-headed	0.94 (0.23)	0.96 (0.19)	0.94 (0.24)	0.91 (0.29)	0.05**	0.03	0.02
Age	Continuous	Age of household head in years	44.72 (12.13)	43.56 (12.01)	44.45 (12.47)	47.47 (11.40)	-3.91***	-3.01**	-0.90
Household size	Continuous	Total Number of household members	4.52 (1.59)	4.72 (1.62)	4.32 (1.55)	4.51 (1.57)	0.21	-0.19	0.41***
Housing status	Continuous	Total number of rooms	1.51 (0.66)	1.58 (0.70)	1.51 (0.65)	1.38 (0.59)	0.20***	0.14*	0.06
Assets	Continuous	Total number of assets	6.34 (5.52)	7.54 (5.80)	4.91 (4.37)	6.57 (6.23)	0.98	-1.66***	2.64***
Farm size	Continuous	Total land area in decimal	106.21 (123.36)	121.67 (135.12)	83.98 (112.13)	116.09 (113.32)	5.58	-32.11***	37.69***
Poultry and Livestock	Continuous	Total number of poultry and livestock	11.52 (11.75)	13.19 (14.20)	10.68 (10.53)	9.76 (7.47)	3.43***	0.92	2.51**
Fisheries Income	Continuous	Annual Income from fisheries and fisheries related sources (in BDT)	1400.48 (3879.79)	1782.47 (5595.12)	1198.10 (2090.31)	1019.90 (1509.27)	762.57	178.20	584.36
Farm Income	Continuous	Annual Income from Agriculture and agricultural related (in BDT)activities	23106.26 (26876.0)	23971.29 (26623.97)	17666.30 (24082.34)	31211.41 (29943.86)	-7240.12**	-13545.11***	6304.99***
Non-Farm Income	Continuous	Annual Income from Non-Agricultural Activities(in BDT)	21530.41 (16282.3)	19606.06 (15947.86)	23800.46 (15292.23)	21196.79 (18177.29)	-1590.73	2603.67	-4194.41***

Variables	Type	Definition and measurement	Total	Participator	Dis-participator	Non-participator	Diff †	Diff ††	Diff †††
CBO Membership	Dummy	=1 if the household head is member of community based organisation (CBO)	0.85 (0.35)	0.98 (0.13)	0.98 (0.14)	0.38 (0.49)	0.61***	0.61***	0.01
Access to extension	Dummy	=1 if have access to GO or NGOs extension	0.94 (0.25)	0.95 (0.22)	0.92 (0.28)	0.94 (0.24)	0.01	-0.03	0.03
Access to irrigation	Dummy	=1 if irrigated crop land last year	0.63 (0.48)	0.65 (0.48)	0.57 (0.50)	0.71 (0.46)	-0.05	-0.14***	0.08*
Credit	Dummy	=1 if able to access credit	0.92 (0.28)	0.92 (0.27)	0.90 (0.30)	0.93 (0.25)	-0.01	-0.04	0.02
Market information	Dummy	=1 if get agricultural market information	0.84 (0.37)	0.87 (0.34)	0.79 (0.41)	0.85 (0.38)	0.03	-0.05	0.08**
Number (Percent) of Sample			570 (100)	234 (41.05)	216 (37.89)	120 (21.05)			

Note: Standard deviations are in parentheses. For identifying differences in mean values, an independent sample t-test was used. ***, **, * indicate that mean values are significantly different at the 1%, 5% and 10% level, respectively. † Difference between participator and non-participators. †† Difference between dis-participators and non-participators. ††† Difference between participator and dis-participators.

Table A.3.2: Multinomial logit analysis of IAA value chain participation with aggregated sample

Variables	Participators	Dis-participators	Dis-participators†
Education	−0.03	−0.07	−0.04
	(0.05)	(0.05)	(0.03)
Gender	0.76	0.37	−0.38
	(0.69)	(0.67)	(0.49)
Age	−0.04***	−0.02	0.02*
	(0.01)	(0.01)	(0.01)
Household size	−0.07	−0.19*	−0.12*
	(0.11)	(0.11)	(0.07)
Asset	0.09**	−0.01	−0.10***
	(0.04)	(0.04)	(0.03)
Farm Size	0.00	0.00	0.00
	(0.00)	(0.00)	(0.00)
Poultry and Livestock	0.00	−0.00	−0.00
	(0.02)	(0.02)	(0.01)
Fisheries Income	0.03	−0.02	−0.05*
	(0.05)	(0.06)	(0.03)
Farm Income	−0.02***	−0.02***	0.00
	(0.01)	(0.01)	(0.01)
Non-Farm Income	−0.02*	0.00	0.02***
	(0.01)	(0.01)	(0.01)
CBO Membership	4.64***	4.63***	−0.01
	(0.57)	(0.56)	(0.73)
Access to extension	0.45	0.35	−0.10
	(0.66)	(0.64)	(0.42)
Access to irrigation	−0.19	−0.07	0.12
	(0.42)	(0.42)	(0.26)
Credit	−0.51	−0.74	−0.23
	(0.67)	(0.66)	(0.38)
Market information	0.62	0.23	−0.39
	(0.52)	(0.50)	(0.32)
Constant	−1.86	−0.92	0.94
	(1.35)	(1.29)	(1.05)
Number of observations	570		
LR chi2(30)	308.58		
Log likelihood	−450.61932		
Pseudo R2	0.2551		
Prob > chi2	0.00		

Note: Note: Categories in 2012 based on 2007 characteristics, non-participator as comparison group. Dis-participators† =for this column participator as comparison group. Standard deviations are in parentheses. * Significance at 10% level. ** Significance at 5% level. *** Significance at 1% level.

Table A.5.1: Construction procedure of the socio-environmental awareness index

Perception on jth Affect	No	Yes				
Affects score (Ij)	0	1				
Rank of the importance of Jth effect on five point scale (Rm)	0	1	2	3	4	5
Rank interpretation (degree of affect)	None	Very low	Low	Medium	High	Very high
Aggregate perception index of ith farmer (APIi)	APIi= ΣΣIj ΣRm where j=1, 2…..24; m=0, 1, 2…5;					
Average perception index of ith farmer (AvPIi)	AvPIi= APIi/N where N=24 (Total number of impacts)					

Source: Adapted from Rahman (2003a).

Table A.4.1: Household well-being measures for IAA value chain participators, non-participators and dis-participators in Bangladesh

Year	Variables		Household Annual			Consumption Frequency						Observation
			Income‡	Expenditures‡	Food expenditures‡	Fish #	Meat #	Egg #	Pulse #	Fruits #	Vegetables¤	
2007	Non-Participator	Mean	76573.97	77806.59	40324.45	4.77	1.1	0.74	2.29	0.27	6.39	147
		Std. Dev.	45516.68	46375.31	14995.01	3.09	1.02	0.88	0.91	0.5	1.17	
	Participators	Mean	65422.67	66975.51	39185.35	6.75	1.15	0.81	2.2	0.41	5.59	510
		Std. Dev.	37900.25	33825.74	13635.09	3.69	1.15	0.86	1.01	0.67	1.33	
		diff	11151.30***	10831.08***	1139.1	-1.98***	-0.06	-0.07	0.08	-0.14***	0.80***	
2009	Non-Participator	Mean	78214.25	77410.76	36333.61	9.87	1.73	0.86	2.24	0.58	5.84	148
		Std. Dev.	63850.36	55125.83	14351.71	5.73	1.63	1.2	1.68	1.12	1.32	
	Participators	Mean	86676.06	82642.78	37834.24	12.68	1.94	0.9	2.41	0.5	5.62	509
		Std. Dev.	50773.34	52895.89	13741.55	5.92	1.9	1.06	1.61	0.94	1.56	
		diff	-8461.81*	-5232.02	-1500.63	-2.81***	-0.21	-0.04	-0.18	0.08	0.22	
2012	Non-Participator	Mean	99987.17	81006.29	51409.3	8.72	1.46	1.33	2.69	0.4	5.3	121
		Std. Dev.	58799.42	42586.67	23390.14	7.37	1.57	1.54	1.85	1.04	1.96	
	Participators	Mean	121807.9	101519.8	50589.32	12.24	2.29	1.25	2.85	0.5	5.8	234
		Std. Dev.	79839.17	73995.7	25980	8.07	2.26	1.5	2.06	1.18	1.94	
	Dis-participators	Mean	90293.65	77970.75	45360.56	8.51	1.45	1.08	2.41	0.31	4.83	216
		Std. Dev.	61598.88	49459.97	20087.43	6.68	1.58	1.35	1.73	1.15	2.22	
	diff	†	-21820.74***	-20513.54***	819.98	-3.52***	-0.83***	0.08	-0.16	-0.11	-0.51**	
		††	9693.51	3035.54	6048.74**	0.21	0.01	0.25	0.28	0.09	0.47*	
		†††	-31514.26***	-23549.08***	-5228.76***	-3.73***	-0.85***	-0.16	-0.44***	-0.19*	-0.97***	

± All are in Bangladeshi Taka (1US $=79.75 in 2012)
No of days in a month
¤ No of days in a week
† difference between participator and non-participators
†† difference between dis-participators and non-participators
††† difference between participator and dis-participators
* Significance at 10% level. ** Significance at 5% level. *** Significance at 1% level.

Table A.4.2: Welfare distribution among the IAA value chain participation categories

Year	IAA participation categories	Statistics	Total household income±	Household total expenditures±	Total household Food expenditures±	Consumption Frequency						Obs.
						Fish#	Meat#	Egg#	Pulse #	Fruits#	Vegetables ¤	
2007	Never-Participate	Mean	76573.97	77806.59	40324.45	4.77	1.1	0.74	2.29	0.27	6.39	147
		Std. Dev.	45516.68	46375.31	14995.01	3.09	1.02	0.88	0.91	0.5	1.17	
	Participate in Production	Mean	79621.59	83969.32	44264.27	6.85	1.31	0.85	2.25	0.42	5.75	216
		Std. Dev.	44912.24	38523.81	15037.2	3.57	1.13	0.87	1.14	0.68	1.31	
	Participate in up and down stream	Mean	54990.82	54490.27	35453.9	6.67	1.04	0.78	2.17	0.4	5.47	294
		Std. Dev.	27521.15	20336.46	11138.36	3.78	1.15	0.86	0.9	0.66	1.34	
	T-Test	Diff_1	21583.15***	23316.32***	4870.55***	-1.90***	0.05	-0.04	0.12	-0.13**	0.92***	
		Diff_2	-3047.62	-6162.73	-3939.82***	-2.08***	-0.21*	-0.11	0.03	-0.15**	0.64***	
2009	Never-Participate	Mean	78214.25	77410.76	36333.61	9.87	1.73	0.86	2.24	0.58	5.84	148
		Std. Dev.	63850.36	55125.83	14351.71	5.73	1.63	1.2	1.68	1.12	1.32	
	Participate in Production	Mean	107889.7	105365.5	42226.55	13.34	2.08	0.92	2.39	0.44	5.59	216
		Std. Dev.	54142.69	57735.1	14221.05	5.96	2.09	1.03	1.66	0.94	1.56	
	Participate in up and down stream	Mean	71037.35	65891.55	34596.22	12.19	1.84	0.88	2.43	0.54	5.65	293
		Std. Dev.	41811.28	41831.38	12442.38	5.86	1.75	1.08	1.59	0.95	1.56	
	T-Test	Diff_1	7176.9	11519.21***	1737.39	-2.32***	-0.11	-0.03	-0.19	0.04	0.2	
		Diff_2	-29675.43***	-27954.75***	-5892.94***	-3.47***	-0.35*	-0.06	-0.15	0.14	0.26	
2012	Never-Participate	Mean	99987.17	81006.29	51409.3	8.72	1.46	1.33	2.69	0.4	5.3	121
		Std. Dev.	58799.42	42586.67	23390.14	7.37	1.57	1.54	1.85	1.04	1.96	
	Participate in Production	Mean	134765.4	120974.6	55384.31	14	2.61	1.48	3.13	0.65	6.03	145
		Std. Dev.	87565.36	84307.05	29031.13	8.14	2.48	1.64	2.2	1.38	1.9	
	Participate in up and down stream	Mean	100697.4	69823.82	42777.25	9.38	1.79	0.87	2.39	0.27	5.43	89
		Std. Dev.	60024.08	35026.02	17560.38	7.11	1.74	1.13	1.72	0.69	1.95	
	T-Test	Diff_1	-710.21	11182.47**	8632.05***	-0.66	-0.32	0.47***	0.3	0.13	-0.13	
		Diff_2	-34778.25***	-39968.33***	-3975.01	-5.28***	-1.14***	-0.15	-0.44*	-0.25*	-0.74***	

Diff_1 = difference between never-Participators and Participate in up and downstream activities
Diff_2 = difference between never-Participators and Participate in production process
± All are in Bangladeshi Taka (1US $=79.75 in 2012)
No of days in a month
¤ No of days in a week
* Significance at 10% level. ** Significance at 5% level. *** Significance at 1% level.

Table A.1.2: Interview Schedule for IAA value chain actors in Bangladesh[41]

**The Interview schedule for the study
on
Impact of Technological Innovation on the Poor: Integrated Aquaculture-Agriculture in Bangladesh**

Abu Hayat Md. Saiful Islam
Junior Researcher / Ph.D. Fellow
Center for Development Research (ZEF)
Bonn, Germany
Mobile: +8801676994359 (In Germany: +4917637734099)
Fax: + 49 228 731839
E-mail: saiful_bau_econ@yahoo.com

....

The Interview schedule

The Interview schedule is designed for an extensive fieldwork in Bangladesh for investigating the impact of technological innovations in agriculture on the rural poor

Sample/farmer code No.:................FFS code...............HH Type..............

Project-1, Non-project-2

Village: _____ Para.................... Union: _____

Upazila: _____ District: _____

Name of the enumerator: _____

Name of the respondent: _____

Name of household head: _____

Religion...................................Ethnic group..

Code: Religion; Muslim-1, Hindu-2, Christen-3, Buddhist-4, others-5 specify **Code: Ethnic:** 1 Santal, 2 Oraon, 3 Pahan, 4 Munda, 5 Mahato, 6 Garo, 7 Hajong, 8 Dalu, 9 Rabidash, 10 Coch, 11 Barman, 12 Mahali, 13 Karmakar, 14 khatrio, 15 Malo, 16 Singh

Typology of farming/technology/activities practiced in 2009; (i)Fish culture in rice field/**(ii)**Fish and aquatic animal production in low land/(iii)Fingerling production in pond/ditch/(iv)Cage culture/(v)Food Fish culture in pond/ditch

If you are not doing now, why?
1.
2.
3.

What you are doing instead of that. ---

Who is responsible for farming decisions (Name): _____

When did you start this farming (Year): _____

Phone/mobile no of the respondent: _____

Relation with household head: _____

(Use relation code given under question no. 01)

_____   _____
Signature of the Supervisor and date   Signature of the Enumerator and date

---

41 This is the merge questionnaire. Originally there were two questionnaires, one for IAA production related value chain actors and other were for up and down stream actors. But most of the questions were similar in both questionnaires except some additional questions for up and down stream actors which are added here.

01. Details household's profile (Include those who take food from the same chula/khana):
Total number of household members: ....

Sl. No.	Name of the member	Relation with the household head (Code-1)	Gender 1 = Male 2 = Fem.	Age (Yrs)	Marital Status (Code-2)	Primary Occupation	Secondary Occupation	Education	Health Status				Member of any organisation/ cooperative and name Code-5	If yes, How many years
						Name of occupation	Name of occupation	Years of schooling	Whether suffered from illness during the last one year 1 = Yes 2 = No	If "yes", what disease?(code-3)	Kind of treatment undergone (Code-4)	Cost of treatment		
1	2	3	4	5	6	7	8	9	12	13	14		16	17

Code-1: (Relation) 1. Household head 2. Wife/husband 3. Son/daughter 4. Father/mother 5. Brother/sister 6. Son/daughter-in-law 7. Grand son/daughter 8. Nephew/niece 9. Brother/sister-in-law 10. Brother's wife/sister's husband 11. Others (specify)

Code-2: (Marital status) 1. Unmarried 2. Married 3. Widow/widower 4. Divorced 5. Separated

Code-3: Cough fever-1, Typhoid-2, Rheumatic fever-3, Diarrhea-4, Dysentery-5, Gastric Ulcer-6, Anemia-7, Asthma-8, Tuberculosis-9, Skin disease-10, Night blind-11, Cancer-12, Pneumonia-13, Chicken pox-14, Blood pressure-15, Mental disorder-16, Paralysis-17, Unknown-18

Code-4: (Kind of treatment undergone) 1. No treatment given 2. Ayurvedic/Quack 3. Homoeopathy 4. Allopathy 5. Government medical centre

Code-5: (Membership of organisation/cooperative) 1. NGO member 2. Member of farmers' cooperative 3. Member of cooperative society 4. Member of Union Council 5. Active member of a political party 6. Others (specify)

02. Migration and social network. If any member migrated. Details of the migrant members living outside the household:

Name of the member	Reason for migration (Purpose) (Code-1)	Informer of work (code-2)	Place of migration (Where) (Code-3)	Type of Migration (Code-4)	Remittance/ money received from the member (last year)	Amount spent for migration last year	What was the source of fund? (Code-5)	Where you invest the remittance money
1	2	3	4	5	6	7	8	9

Code-1: (Purpose) 1=Job 2=on a visit 3=Medical treatment 4=Study 5=Wage-labour 6=Business 7=Others (specify)

Code-2: (Informer of work) 1=Member of Family, 2=Uncle/Aunt, 3=Relative of Father-in-law, 4=Member of society/Neighbor, 5=Chairman/Elite person, 6=Broker/Sarder, 7=Political Leader/Worker, 8=Big service holder/Industrial/Businessman, 9=other (specify)

Code-3: (Where?) 1=Within this district 2=Another district within the country 3=Foreign country

Code 4: (types) 1=domestic temporary, 2=domestic permanent, 3=international permanent migration.

Code-5: (Source of fund) 1=from own savings 2=by taking loan 3=by selling land 4=By mortgaging property 5=By taking help from relatives 6=By selling property

2.1 Which months of the year are seasonal wage earners typically working away from the household?

J	F	M	A	M	J	J	A	S	O	N	D

2.2 Do you have other extended family members living in your village or city neighborhood? Yes -1, No-0

2.3 How often does your household receive or give cash or in-kind support? 0. Not at all 1. Only at festivals 2. All year round

2.4 Is anyone in your family a member of the following decision making bodies in your community? 1. Mosjid or mondir or church committe 2. Local school college, madrasha 3. Union parishad 4. Upazila or MP

3. Do either relatives, friends or neighbors help this household in any of the following ways, or does your household help others? (put tick mark)

Type of assistance	Help to household		Help from household	
	Relatives	Friends or neighbor	Relatives	Friends or neighbor
General support				
Finding a job				
Help to pay education and medical expenses				
Help with rent and housing costs				
Help in paying for agricultural inputs				
Help paying off debts				
Lending money or in kind goods				
Contribute to wedding or engagement expenses				
Help with looking after children				
Provide labour to assist household				
Sharing income generating equipment				

4. Land resources (last year)

How is land measured locally? (specify) .... equivalent to decimals............How many **plot cultivated last year**..................**quantity (dec)**..................**Irrigated area**..................

Crop Plot name	Total land (decimal)	Own (individual)	Own (not yet distributed)	Private	Public (Khas)	Leased out (dec)	Leased in (dec)	Rent-in	Rent-out	Mortgage in	Mortgage out
Rice–fish plot											
Pond											
Dyke (pond)											
Cultivable land (crop/vegetable) excluding rice–fish											
Homestead area											
Homestead vegetables/fruits garden											
Bamboo/wood garden											
Fallow land (not arable)											
Others											

4.1 How many decimals of land is sold in last 10 years?..
Decimals..........................Why..
4.2 How many decimals of land is acquired in last 10 years?...
Decimals..........................How..

5. Non-land assets.
5.1 Housing status

SL No of Room	Type of the dwelling house-code			Present value in Taka	Last year investment in any of this housing
	Wall	Roof	Floor		

Code:

Wall	Earth-1	Bamboo/wood-2	Brick-3	Jute stick-4	Straw-5	Tin-6	Others-7
Roof	Straw and shon-1	Brick-2	Tally-3	Tin-4	Leafs-5	Others-6	
floor	Earth-1	Brick/concrete-2	Others-3				

5.2 What is your main source of **lighting**?
5.3 What is your main source of **cooking fuel**?
5.4 What is the **source of your drinking water**?
Shallow tubewell / deep tubewell / bottle water / pond/river/lake water / hand tube well /reservoir/fountain water/others (specify)
5.4.1 **Ownership in case of tube well** and/or pond or others: **Own / Shared / Government / NGO**
5.4.2. Do you use **purifier/boil water before drinking**? (yes=1, No=0)
5.4.3. Is the water that you are drinking free **from arsenic contamination**?(Yes=1, No=2, Don't know=3)
5.4.4. If the answer is yes/no, how do you know that water is free/not free from arsenic?......................
5.4.5. What you do if water is arsenic contaminated?
5.5 **Sanitation (type of latrine)**: Brick latrine / Ring Slap/sanitary latrine / *Katcha* latrine /bamboo/palm leaf made open latrine/ Open field
5.5.1 **Ownership of latrine**: Own / Shared / Neighbors / Others
5.6 **Do** you have **telephone in your hoseholds**? (Land or mobile or both) Yes......... 1, No.........0 /___/

5.7. How much did you pay last month for?

Item Name	Amount paid	Item Name	Amount paid
Electricity		Trash collection	
Coal		Solar energy (Taka and WT)	
Kerosene		Telephone/mobile	
Wood (purchased)		Water	

6. Household assets (other than land):

Names / Types of assets	No.	Value in Taka	Who is the owner (M/F)
6.1 Income assets:			
6.1.1 Agricultural income assets			
Cow			
Buffalo/bullock			
Goat			
Sheep			
Pig/swine			
Chicken			
Duck			
Pigeon			
Tiber trees (Mahogani, bamboo etc.)			
Fruits trees (Mango, Jackfruits, brittle nut, coconut etc.)			
Tree nursery			
Shallow machine			
Hand tube well			
Power tiller/tractor			
Other agricultural instruments (plough, harrow, weeding tool, etc)			
Fishing net			
Husking machine			
Sugarcane crusher			
Threshing machine			
Spray			
Bullock cart			
Others (Please. Specify):			
6.1.2 Non-Agricultural income assets			
Business shop			
Sewing machine			
Rickshaw/ van			
Boat			
Motorcycle			
Tempo/Nasiman /Auto			
Mobile phone			
Money lending (*Dadon* capital)			
Others (Please. Specify):			
6.1.3 Non-income household assets			
Chair			
Table			
Sofa set			
Almira			
Alna			
Cot (khat) and chouki			
clock			
TV /radio/VCD			
Freeze			

Bicycle			
Jewelry			
Gas stove or electric stove or wood stove			
Washing machine			
Personal computer			
Car			
Others (Please. Specify):			

7. Sources of income

7.1 a) **Total household annual income**...................................Tk.
b) **Farm income** (Income from on-farm activities per year)..................
c) **Off-farm income** (Income from off-farm sources per year).......................
d) **Non-farm income** (Income from non-farm sources per year)............................
e) **Income from fisheries /aquaculture activities**..........................Income from rice.....................

7.2 Agricultural income (all IRF/pond fish/cage and land income)

What are your household's income generating activities in order of importance?	Total farm produce (Kg or liter or number)	Gross income (Tk.)	Consumption (Kg/year)	Sales (Kg/year)	Price per Kg.	Who participates in this activity and control money? (use code-12)
Fish harvest (wild fish)						
Fish harvest (Kutchia, oyster etc)						
Fish trading						
Fingerlings trading						
Fish processing for sale						
Netting team (self)						
Fish net/gear making						
Fish /prawn Hatchery						
Fish fry, crab frog and Dry fish						
Other, specify						
Fish culture						
Rice						
Wheat						
Maize						
Jute						
Sugarcane						
Pulse						
oilseed						
Vegetables						
Potato						
Nursery (fruit, timber etc)						
Fruits						
Coconut and betel nut (Tk)						
Condiment and Spices						
Others						
NON-CROP ACTIVITIES (Livestock and poultry)						
Cow or oxen						
Goat						
Sheep						
Swine						
Buffalo						
Duck						
Beef						

Mutton						
Milk						
Poultry birds						
Eggs (no.)						
Animal skins						
Cow dung						
Forest Product						
Bamboo						
Fruit and Timber Trees						
Firewood trees						
Honey						
Selling seed						
Flower sale						
Other, specify						

Participant and control codes-12; 1 Men only, 2 Women only, 3 Children only, 4 Adults only, 5 Women & children, 6 Men & children, 7 Everybody, 8 men and women shared

7.3 Non-farm and off farm income sources

What are your household's income generating activities in order of importance?	No of HH member involved	Last year Income (Tk/year)	Who participates in this activity? (use code)
Govt. Services of farmer himself & household members			
Private services of farmer himself & household members			
Labor selling for Ag. (farmer himself & household members)			
Labor selling for Non-Ag. (farmer himself & household members) please specify	1.		
	2.		
	3.		
Business (medium and large scale)			
Small/petty trading / small grocery shop keeping			
Tempo/van/rickshaw /motorcycle renting			
Tempo/van/rickshaw /motorcycle/thela gari driving			
Shallow pump rented out			
Power tiller and/or plough renting			
Fishing net renting			
Remittance (in country and abroad)			
Money lending			
Land leased and/or mortgage out			
Navigating boat			
Handicrafts			
Carpet weaving			
Taxi/transport or Driving car			
Pension and other Govt. benefits (eg. Freedom fighter allowance)			
Sale of food aid			
Begging/borrowing Other			
Cottage industry/Industrial labour			
Construction and repair of houses and roads			
Catching fish in khals & beels			
Collecting fuel wood/thatch etc. from forest			
Others (Please specify)			
Donation from other community			

Participant and control codes-12; 1 Men only, 2 Women only, 3 Children only, 4 Adults only, 5 Women & children, 6 Men & children, 7 Everybody, 8 Men and women shared

8. Household annual Expenditures

Items	Annual Expenditures (Tk.)	Items	Annual Expenditures (Tk.)
Food		Loan repayment	
Clothing		Land purchase	
Education		Land rent	
Health		Livestock/poultry	
House repair/building		Cost of inputs	
Festivals, ceremonies. marriage		Furniture purchase	
Farm equipment		Savings	

9. Household food security level

9.1 Did you face any lean period in taking sufficient food? Yes / No *(Please, put ✓)*

9.1.1 Food supply status of the HH in last year

Month	Crises month (Yes-1, No-0)	No of meal per day
Baishakh (april-may)		
Jaistha (may-jun)		
Ashar (Jun-Jul)		
Shraban (Jul-Aug)		
Vadra (Aug-Sep)		
Ashwin (Sep-Oct)		
Kartic (Oct-Nov)		
Aghrayan (Nov-Dec)		
Poush (Dec-Jan)		
Magh (Jan-Feb)		
Falgun (FEB-Mar)		
Chaitra (Mar-April)		

9.3 Households self-assessed in food security status last year (Food deficit-1, Breakeven-2, surplus-3)…

9.4 Food security situation in this year compare to 2009 (Improved-1, same-2, Worsen-3)……………

9.5 Recall for last week/month food consumption by the household

Food items		Number of days	Total meal or number per week/month	Consumption per day
Cereals	Last week			
Fish	Last month			
Meat	Last month			
Egg	Last week			
Pulses	Last week			
Milk	Last week			
Fruits	Last week			
Vegetables	Last week			

9.6 Which salt you eat normally – open or packet or iodine rich salt………………

9.7 Household Food Insecurity Access Scale (HFIAS)

NO	Question	Response
1	In the past four weeks, did you worry that your household would not have enough food?	(Yes=1 No=0)
1a	How often did this happen? 1 = Rarely (once or twice in the past four weeks) 2 = Sometimes (three to ten times in the past four weeks) 3 = Often (more than ten times in the past four weeks)	

2	In the past four weeks, were you or any household member not able to eat the kinds of foods you preferred because of a lack of resources?	(Yes=1 No=0)
2a	How often did this happen?	
3	In the past four weeks, did you or any household member have to eat a limited variety of foods due to a lack of resources?	(Yes=1 No=0)
3a	How often did this happen?	
4	In the past four weeks, did you or any household member have to eat some foods that you really did not want to eat because of a lack of resources toobtain other types of food?	(Yes=1 No=0)
4a	How often did this happen?	
5	In the past four weeks, did you or any household member have to eat a smaller meal than you felt you needed because there was not enough food?	(Yes=1 No=0)
5a	How often did this happen?	
6	In the past four weeks, did you or any other household member have to eat fewer meals in a day because there was not enough food?	(Yes=1 No=0)
6a	How often did this happen?	
7	In the past four weeks, was there ever no food to eat of any kind in your household because of lack of resources to get food?	(Yes=1 No=0)
7a	How often did this happen?	
8	In the past four weeks, did you or any household member go to sleep at night hungry because there was not enough food?	(Yes=1 No=0)
8a	How often did this happen?	
9	In the past four weeks, did you or any household member go a whole day and night without eating anything because there was not enough food?	(Yes=1 No=0)
9a	How often did this happen?	

9.8.2 Self-assessment of total household health status compared to 2009. (Improved-1, same-2, worsened-3)
9.8.3. Self-reported assessments of respondents' ability to perform daily activities
Ranging from walking one kilometer..
Lifting heavy objects ....
Simple activities such as eating, bathing, or dressing..

10. Contract with Extension and training and reason for farming

Items	Male (frequency in the last three month)			Female (ask the female member if possible)		
	Response (yes = 1, 0 for otherwise)	Frequency of visit to your HH	Frequency of visit by you	Response (yes = 1, 0 for otherwise)	Frequency of visit to your HH	Frequency of visit by you
Do you have any contact with Go or NGO extension						
Upazila fisheries officer						
Agricultural office						
Livestock office						
Land office						
Family planning						
Cooperative office						
Barendra project						
Union parishad office						
NGOs						
Donor						
Do you have experience with the following						

Method demonstration (crop, livestock and fish)							
Result demonstration (crop, fish, livestock)							
Group discussion							
FFS training program							
Other training program (specify)							
Watch agricultural poster							
Reading agricultural books and magazines							
Listening agricultural related programs in Radio							
Watch agric related program in Television							
Participating national program (fish fair, agricultural fair)							
Farmers rally							

Had access to seed(Y or N)	
Had access to feed(Y or N)	
Has information on farm technology(Y or N)	
Had marketing information(Y or N)	
Do you get any training on pond fish or Cage or prawn or rice or IRF farming(Y or N)	
If yes then duration	
who got the training	
how many members from the family	
Source of training	
How fish farming experience acquired 1. Self-study 2. Friend & neighbours 3. NGO 4. Demonstration plot 5. Other (please specify)	
Do you think that this type of farming is more profitable than other? (Y or N)	
Do you think fish/cage/RF farming is more relax than agriculture? (Y or N)	
If you have more land, do you like to convert into pond or integrated? (Y or N)	

10.1 Farmers' perception on temperature and precipitation changes during last 20 years.

Farmers' perception	Increase	decrease	Stayed the same	What are the adaptation strategies (code)	Why you did not adapt (barriers)? (code)
Overall Annual Temperature					
Dry season temperature					
Rainy season temperature					
Length of cold periods					
Length of hot periods					
Overall Annual Precipitation					
Rainy season					
Dry season					
Length of dry spells, rainy season					
Intensity of rainfall events					
Inundation of fields and villages					

Code: Farmers' adaptation strategies:- 1=No adaptation, 2=Different crops/mixed cropping 3= Use of drought/ early maturing crop varieties 4=Use different varieties 5=Change planting dates (early or late) 6=Change from one crop to another 7=Migrate to urban areas 8=Change from crops to livestock 9=Change from livestock to crops 10= Non-farming to farming 11= Find off-farm job 12= Build a water harvesting scheme 13= Buy insurance 14=Increase water conservation 15=Put trees for shading and shelter 16=Use more irrigation 17=Implement soil conservation techniques 18=No adaptation 19=Other adaptations (specify...

Code: Barriers to adaptation 1= changes are expensive 2= lack of markets 3= no access to water 4= Lack of knowledge on adaptation 5= Lack of improved seed 6= Rationing of inputs 7= lack of information on changes in the climate (temp and rainfall) 8= Lack of credit/money 9=Lack of information on appropriate adaptation options 10= shortage of land 11= Insecure property rights 12= shortage of labour, 13=Poor potential for irrigation, 14=No barriers, 15=others (specify)

11. Characteristics and utilization of the parcels cultivated by the household (period 2011/12). How many parcels of land in total? ……………………

Sl. No. of the parcel	Identity of the parcel	In whose name the land is registered (Code-2)	Amount of land (in decimals)	Soil type (Code)	Land elevation (depth of flooding during the peak of the monsoon season) (Code)	Source of Irrigation and pump ownership (Code)	If rented-in or out, tenurial arrangement (how used) (Code)	Utilization of land									
								Boro/Rabi Season				Aus Season				Aman Season			
								Crop grown and management (Code)	Name of the variety (Code)	Total harvest		Crop grown and management (Code)	name of the variety (Code)	Total harvest		Crop grown and management (Code)	name of the variety (Code)	Total harvest	
										Maund	Pr			Maund	Pr			Maund	Pr
1	2	3	4	5	6	7	8	9	10		11	12	13		14	15	16		

Code - 2: (Land ownership) 1=Husband 2= Wife 3=Father 4=Mother 5=Another member of the family * Write more than one (corresponding) code, in case of joint ownership.
Code - 4: (Soil type) 1=loamy 2=sandy-loam 3=clay 4=clay-loam 5=sandy
Code - 5: (Land elevation) 1=High-land (very little standing water) 2=Medium-land (knee-deep water) 3=Low-land (Chest-deep water) 4=Very low-land (water more than chest height)
Code - 6: (Source of irrigation) 1=Low-lift pump (L.L.P.) 2=Shallow tube-well (S.T.W.) 3=Deep tube-well (D.T.W.) 4=Irrigation canal project of Govt. (WDB) 5=river/pond/tank Local irrigation system 6=Without irrigation; if Pump irrigation then ownership: self=1, other Adivasi community member=2, other non Adivasi member=3, Adivasi cominuity organisation=4, Non Adivasi community organisation=5, Barendra project=6, shared with Adivasi=7, shared with non Adivasi=8, others=9
Code - 7: (Tenurial arrangement) 1=Self cultivated 2=Sharecropped-in 3= Sharecropped-out 4=Leased-in 5= Leased-out 6=Mortgaged-in 7= Mortgaged-out 8=Others (specify)
Code - (8)(11)(14): (Crop cultivated) *=for rice, prawn and fish write the names; 1=Wheat 2=Jute 3=Sugarcane 4=Oil-seed 5=Pulse 6=Potato 7=Onion 8=Spices 9=Vegetables 10=Mixed Rabi crops 11=Others (specify)
Code - (9)(12)(15): (Crop variety of rice) 1=Local 2=Modern/HYV 3=Hybrid, for fish species write the species name. Management: best management practice (BMP)-1, o otherwise.

12. Information about one plot – write the serial no from previous table……

12.1 Why you doing this?
1. Positive perception about the profitability (yes or no) if yes then level (0 to 5 scale……………..)
2. Positive perception about its environmental effect (yes or no) if yes then level (0 to 5 scale……………..)
3. Positive perception about its nutritional diversity effect (yes or no) if yes then level (0 to 5 scale……………..)
4. Positive perception on adaptability to the biophysical condition (yes or no) if yes then level (0 to 5 scale……………..)
5. Others

12.2 Why not doing this? (Barriers/constraints)
12.3 Crop calendar of the respective plot/activity

Month	Activity
Baishakh (april-may)	
Jaistha (may-jun)	
Ashar (Jun-Jul)	
Shraban (Jul-Aug)	
Vadra (Aug-Sep)	
Ashwin (Sep-Oct)	
Kartic (Oct-Nov)	
Aghrayan (Nov-Dec)	
Poush (Dec-Jan)	
Magh (Jan-Feb)	
Falgun (FEB-Mar)	
Chaitra (Mar-April)	
Other if any	

12.4 Cost and return of the practice-rice only or IRF or IRP or pond fish or cage culture or others specify…………………………….. and for how many decimal………………………..…………

Sl. No.	Particulars	Boro/Rabi Season Boro	Aus Season/vegetables	Aman Season
1	2	3	4	5
Crop				
1	Name of the crop			
2	Crop or fish or prawn or dike crop variety[42]			
Land preparation				
3	How many times have you tilled the land?			
4	If plough was hired, then its expenditure (Tk.)			
5	If mechanical plough was used, then its expenditure (Tk.)			
6	Family labour[43] in ploughing (days)			
7	Hired labour (days)			
8	Daily wage (Tk.)			

---

42 Write the name of the variety or species like ….BR28 or Bagda..golda…etc.
43 In case of labour always mention if male-M and if Female-F.

Sl. No.	Particulars	Boro/Rabi Season Boro	Aus Season/vegetables	Aman Season
1	2	3	4	5
	Seed or seedling cost			
9	Amount of seed (Kg.) and cost			
	Land preparation (ploughing) No and cost			
	Fertilizers UREA (amount and price per Kg)			
	TSP (amount and price per Kg)			
	DAP (amount and price per Kg)			
	GIB fertilizer (amount and price per Kg)			
	Cow dung (amount and price per Kg)			
	others			
	Pesticide cost for seedlings (No., amount and price)			
	Family labour for seed, fertilizer and pesticide application			
	Hired labour for seed, fertilizer and pesticide application and wage rate			
10	Cost of seedling if purchased (Tk.)			
	Family labour for seedling pulling out (M-days)			
	Hired labour for seedling pulling out (M-days)			
11	Family labour for crop establishment (M-days)			
12	Hired labour (M-days)			
13	Daily wage (Tk.)			
	Irrigation cost			
14	Age of seedling.			
15	How many seedlings plant in a bunch			
	Fish fingerlings by species			
	Quantity			
	Price or cost			
	Quantity			
	Price or cost			
	Quantity			
	Price or cost			
	Quantity			
	Price or cost			
	Average size of the fish fingerlings			
	Equipment Cost for vegetables cultivation			

Sl. No.	Particulars	Boro/Rabi Season Boro	Aus Season/vegetables	Aman Season
1	2	3	4	5
Dike/pond preparation				
	Machine cost for pond drying/and pack clear			
	Number of family labour			
	Hired labour			
	Wage rate			
	Lime quantity and cost			
	Others			
Fertilizer				
16	Amount of Urea (Kg.) and type[44]			
17	Price per Kg or Cost of Urea (Tk.)			
18	Amount of Phosphate (TSP) (Kg.)			
19	Price per Kg or Cost of Phosphate (Tk.)			
	DAP(amount Kg)			
	Price per Kg or Cost of DAP			
20	Amount of dosta			
	Price per Kg or Cost of dosta			
	Amount of MP (Kg.)			
	Price per Kg or Cost of MP (Tk.)			
	Cost of other chemical fertilizers (Tk.)			
21	Amount and Cost of cow-dung/green manure (Tk.)			
	others			
	Family labour for fertilizer application (days)			
	Hired labour for fertilizer application days)			
	Daily wage (Tk./day)			
Irrigation				
22	Source of irrigation (Code)			
23	No and Irrigation/fuel cost (Tk.)			
Weeding				
24	Number of times weeded			
25	Family labour (days)			
26	Hired labour (days)			
27	Daily wage (Tk.)			
Pesticide				
28	Amount of pesticide			
	Cost of pesticide (Tk.) and Name			
29	Number of times applied			
30	Family labour for application (No.)			
	Hired labour for application (No.)			

---

44 Write if Guti Urea-1, 0 otherwise.

Sl. No.	Particulars	Boro/Rabi Season Boro	Aus Season/vegetables	Aman Season
1	2	3	4	5
	Wage rate (Tk/day)			
	Lease value of the land (Tk/ season or Tk./ year)			
	Feed (quantity and price per unit)			
	Urea(Q)			
	Price per unit			
	Cow-dung (Q)			
	Price per unit			
	Fish meal (dryfish bran) (Q)			
	Price per unit			
	Packet/Pillet feed (Q)			
	Price per unit			
	Boiled rice(Q)			
	Price per unit			
	Snail as feed (Q)			
	Price per unit			
	Rice or wheat bran(Q)			
	Price per unit			
	Rice or wheat flour(Q)			
	Price per unit			
	Homemade feed(Q)			
	Price per unit			
	Mastered oil cake (MOC) (Q)			
	Price per unit			
	Gas tablet (cost)			
	Bleaching powder (cost)			
	Rotanon/Deptares (cost)			
	others			
	Family labour for application (M-day)			
	Hired lobour for application (M-day)			
	Wage rate (Tk/day)			
	Harvesting			
31	Family labour for harvesting (days)			
32	Hired labour (days)			
33	Daily wage (Tk.)			
	Threshing and cleaning			
34	Family labour for threshing (days)			
35	Hired labour (days)			
36	Daily wage (Tk.)			
	Or Total cost of threshing			
	Selling and buying of input and output			
	Family labour (days)			
	Hired labour (days)			
	Daily wage (Tk.)			
	Transport and other cost if any			

Sl. No.	Particulars	Boro/Rabi Season Boro	Aus Season/vegetables	Aman Season
1	2	3	4	5
	Other expenditures, if any			
37	Family labour for removing crop residue (days)			
	Hired labour (days)			
	Daily wage (Tk.)			
	Amount of other expenditures, if any, for cultivation (Tk.)			
	Output			
38	Crop produced (40kg=1mound)			
	Crop sell			
	Own consumption and gift			
39	Crop Price (Tk.)			
40	Amount and Market price of by-product. (Tk.)			
	Write fish species produced including own consumption name with quantity and price			
	Indigenous fish			
	Write Vegetables species produced including own consumption name with quantity and price			
	Crop damage			
42	Reasons of crop-damage t 1=Pest attack; 2=Flood; 3=Drought; 4=Hailstorm; 5=Others			
43	Amount of crop-loss (Maund)			
44	In case of pest attack or diseases, names of the insects/diseases			

12.5 Did you test the soil before cultivation……………..yes or no and Water test……………..yes or no

13. Investment cost for pond or IRF or cage or others enterprise

How much money did you spend for gher/pond construction at the beginning of prawn/fish farming:....................year....................

Items	Number (Q)	Own				If rented, total cost of renting(Tk/year)
		Purchase value/cost (Tk)	Purchasing year	Durability (life span) (year)	% used for selected *enterprise*	
Bamboo/wood/rope						
Shallow tubewell/pump						
Spade/cycle etc,.						
Drum/box/fishing trap						
Boat/tube						
Net (harvesting)						
Blue net (Hapa and fence)						
Gher/pond house						
Others						
Tools repairing (housing, net, pump) Tk/year						
Salary of management Staff (Tk/yr)						

14. Buying and selling of input and output[45]

Input and output	Source of buying/ lease or sell	%	Cash or credit or contract	How price is determined and who play the key role
Rice and other crop seed	1.			
	2.			
	3.			
Fish fingerlings	1.			
	2.			
	3.			
Fish/prawn feed	1.			
	2.			
	3.			
Fertilizer and pesticides	1.			
	2.			
	3.			
Rice output and byproduct	1.			
	2.			
	3.			
White fish, Crab (*Kakra*), Shrimp/prawn, Kuthcia	1.			
	2.			
	3.			

---

45 In case of multiple sources please mention share in percent (%).

Other outputs	1.			
	2.			
	3.			
What are the value addition work you do for your input and output (like, cleaning, grading, icing etc)	Input: Output:			

15. **Buying and selling of the respective product (for up and downstream actors only)**

How many month run this business in a year..................................

Input and output name	Buying sources	%	Price per unit	Vale addition functions	Selling sources	%	Price per unit	How price determined	Who play the key role	Annual sells or purchase volume

16. **Cost and return for running the respective input or output/product business/activity (for up and downstream actors only)**

Operational /variable Cost items	Unit	Quantity	Price per unit	Total cost last year
Money investment/ running capital				

16.1 Investment/fixed cost (for up and downstream actors only)

Equipments	Qnt.	Own			If rented,
		Purchasing price	Purchasing year	Expecting life time (year)	total cost of renting(Tk/mn)

16.2 Output/Income (for up and downstream actors only)

Items	Daily average purchase and sells volume	Annual purchase and sells volume (Q)	Annual average buying price	Annual average selling price

17. Do you have access to credit or loan / credit? Yes or No………..
17.1 Did you take the credit or loan? Yes or no………. Repaying loan or credit Yes or No…….
17.2 If yes, fill table below and response your degree of loan repayment ability (Highly capable-1, Capable-2, Not capable-3)………….

Name of source	Purpose of loan	Cash or kind	Total amount taken (Taka)	Total amount of repayment (Taka)	Duration of repayment (Months)	Interest rate

18. Tenure security and Soil quality

Items	Response (yes = 1, 0 for otherwise)
Do you expecting a reduction in land size over the coming five years	
Have you been asked to leave your land since renting or leasing?	
Have you ever had a dispute over the ownership or leasing period of this land?-	
Soil of the respective plot	
Plots slope (steep=1, Medium=0)	
Soil quality (Infertile/poor-0, fertile/good (maintained)=1, Highly fertile/ very Good=2,)	
Relative land size (household's land size divided by village average land size) (big-1 small-0)	
Subjective risk aversion capacity-do you love to take the risk	
if yes at what percentage	
Presence of wetland/low land area (present = 1, not present = 0)	
Topography (% of parcels) (Flat, Gentle slope, steep sloping, terraced, contour bunds, eroded terrain Others)	
Water color (Clear-1, Muddy-2, Greenish-3)	
Presence of water plants (Plenty-1, Moderate-2, Trace-3)	
Average water depth, Winter……………..ft and Summer ………….…..ft	
Use of water (Fish culture only-1, Fish culture and household use-2)	

19. Farmers' perception on beneficial and harmful effect of pesticide use

Beneficial effect	Yes (1) or no (0)[46]	Severity[47] (1 to 5 scale)	Harmful effect	Yes (1) or no (0)	Severity (1 to 5 scale)
Destroy insects			Do not affect much		
Increases production			Damages plants if used in excess		
More fertilizer more production			Affects human health		
Increase soil fertility			Cause fish destruction		
Prevents disease infestation			Cause death of livestock/poultry		
Good plant growth			Tasteless product		
Require less fertilizer			Production falls if used in excess		
			The quantity of beneficial organisms (eg. Earthworm, frog) and useful insects will decrease		
others			Destroys soil fertility		
			Pollutes water body		
			Pollutes air		
			others		

19.1 Farmers' perception on the effect of different farming systems and technological change within these farming systems (like environmental, economic, socio-cultural)

Effect	RM		IRFP/IRF		Only fish culture in pond	
	Yes (1) or no (0)[48]	Severity[49] (1 to 5 scale)	Yes (1) or no (0)	Severity (1 to 5 scale)	Yes (1) or no (0)	Severity (1 to 5 scale)
Affects human and animal health						
Reduces soil fertility						
Reduces fish catch						
Increases disease in crops or fish/prawn						
Compacts/hardens soil						
Increases insect/pest attack						
Reduces soil organic matter						
Increases soil acidification						
Increases soil erosion						
Increases soil salinity						
Biodiversity decreases						
The quantity of beneficial organisms (eg. Earthworm, frog) and useful insects are being decreasing						

---

46 A value of 1 is assigned for each of the impact indicators where the farmer recognizes the impact and 0 otherwise.
47 A score of 1 is assigned for least severity and 5 for very high severity.
48 A value of 1 is assigned for each of the impact indicators where the farmer recognizes the impact and 0 otherwise.
49 A score of 1 is assigned for least severity and 5 for very high severity.

Contaminates water source						
Depletion of water table						
Creates water logging						
increase need for inputs						
Production decreasing over the years						
Inefficient fertilization						
Labor demand decreasing						
The role of women in farming activities is decreasing						
Increased use of child labour						
Variety of nutrition consumption decreases						
Conflict among villagers for the issue of irrigation water						
Conflict among villagers for the issue of land						

20. Access to information about input and output prices, technology, best management practice, jobs, government information, news, health etc

Types of information	Do you have access?	If yes then source of information (institution) (Code)
Fish/Prawn and rice sale price		
Fish/Prawn and rice seed price and quality		
Technologies (BMP, CST, new feed, new seed and fish species, new management technique etc)		
Use of medicine		
Drought, flood tolerant varieties		
Water saving technologies		

Sources: 1. Relatives, friends and neighbours, 2. Community bulletin board, 3. Local market, 4. Mullahs, 5. Local newspaper, 6. National newspaper, 7. Groups or associations, 8. Business or work associates, 9. Political associates, 10. Community leaders, 11. An agent of the government, 12. NGO's, 13. Internet, 14. Radio, 15. Television 16. Others specify

21. Major problems faced in farming and the households in the last year

Rank	Name of Problem	Types of losses	Total losses (Tk)	How did you solve the problem
1				
2				
3				

(Ranking: 1=Most critical problem, 2=Second most critical problem, 3=Third most critical problem,....so on)

22. Infrastructure

Infrastructure	Distance from home (Km)	Travel time (minute)	Cost of transportation (Tk)	Main transportation medium	Road condition Good or bad (1, 0)
Pond/IRF/cage/plot					
Primary market (small market)					
Secondary market (big market)					
Storage or preservation facility					
Rice mill					
Paved road					
Bus stop					

Bank						
Union office						
Agricultural extension office						
High school						
College						
Thana (sub-district) headquarter;						
Post office						
Hospital and medicine						
NGO						
Capital (Dhaka)						
Nearest road						
Telephone booth						
Internet booth						
Plot (farming unit)						
Number of bepari/faria in this village						
Number of formal and informal finance sources						

22.1 Institutional linkages

Name of the community organisation (eg, CBO, collective action etc.)	Membership (formal or informal)	Position		Who makes the rule?	How the rule makes?
Common property	access right (yes or no)	If use any problem faced or if don't use then why			
Khas pond					
River					
Beel/reservoir					
Forest					
Grazing land					
Floodplain					

23. Farmers' satisfaction with PF/RM/RF/Cage/respective value chain activity
1—strongly satisfied, 2—satisfied, 3—undecided, 4—dissatisfied, 5—strongly dissatisfied
23.1.1 If satisfied then level of satisfaction, 1 to 10 scale? ....
Compare to other production/livelihood option are you satisfied (yes or no) if yes how much you satisfied (%)?
23.1.2 How much you achieved the expectation of earnings from pond fish/ cage/RFP/RM/RF (%)?

24. Subjective Happiness Scale
Instructions: For each of the following statements and/or questions, please circle the point on the scale that you feel is most appropriate in describing you.
1. In general, I consider myself:

1	2	3	4	5	6	7
not a very happy person						a very happy person

2. Compared to most of my peers, I consider myself:

1	2	3	4	5	6	7
not a very happy person						a very happy person

3. Some people are generally very happy. They enjoy life regardless of what is going on, getting the most out of everything. To what extent does this characterization describe you?

1	2	3	4	5	6	7
not a very happy person						a very happy person

4. Some people are generally not very happy. Although they are not depressed, they never seem as happy as they might be. To what extend does this characterization describe you?

1	2	3	4	5	6	7
not a very happy person						a very happy person

25. Satisfaction and other situation

Are you satisfied with	YES=1, 0= otherwise	If yes then Level-1 to 7 scale
Your income?		
With your life?		
With job		
Financial situation		
Your health status		
Child education		
Your family member health status		
Others		
How satisfied are you with your life		
The amount of smiling in the questionnaire		
Changes in facial muscles		
Individuals' self-reported disabilities or incapacity to perform daily activities		
Self-reported chronic illnesses		
The number of visits to the doctor		
Days staying at hospital		
Having access to health care		
Obesity (body mass index)		
What makes you happy?		

26. What change has taken place in economic condition of the household during last 5 years?

Has improved   No change   Has worsened

Three main reasons of improvement:

1..

2..

3..

Three main reasons of worsening:

1..

2..

3..

27.1 In your own assessment in which category would you put your household, compared to the conditions of other households of your village?

| 1 | Rich | | 2 | Solvent | | 3 | Self-sufficient | | 4 | Poor | | 5 | Very Poor |

And why-the reasons   1.
                       2.
                       3.

28. Woman empowerment

Contribution/participate in Decision making	D Code	E-D-code
Purchase, sell or mortgage of land		
Food grain cultivation		
Cash crop cultivation		
Livestock and poultry rearing		
Homestead production of vegetables and fruits		
Other non-farm activities		
Fish or prawn cultivation		
Cage management		
Family farm management		
Repairing the house/ constructing new house		
Varity or species/crop selection		
Daily HH expense		
Agricultural input expenses		
Child education and expenses (sending to school and keep going)		
Spending own money income		
Spending for charity or help		
Buying or selling HH furniture's		
Helping relatives of parent HH (money or materials)		
Buying and selling of agricultural input and output		
Family planning or child birth.		
Medical treatment and expenses (sending family member to health center and doctors etc)		
Borrowing or giving loan/ money		
Joining with any organisation		
Going to your father house		
Child marriage		
Guest invitation and entertainment		
Buying and selling or transfer of assets		
Local socio-political-cultural and religious meeting/shalish		
Participation or celebration in social functions such as marriage, birthday, invitation, religious festivals		
Helping neighbor's in emergency such as delivery, death, dieses etc		
Work with people to fight situation like tornado, fire, flood etc		
Participation in (Adivasi) community meeting		
Cultural programs (drama, folk songs)		
Other voluntary social program like tree plantation, cleanliness drive etcc		
Arbitration in family quarrel of neighbor and relatives		
Participation in election (casting vote)		
Member of different groups or NGO or any social, cultural, religion organisation (write name and position)		
Contact with different external people from different organisation in the last month (write the org name)		

Do you think you have enough freedom to choose whatever you like	Yes or no	
Have you attended the FFS training program	Yes or no. if yes then number	
If no why?		
Have you attended any other training program if yes name or the organisation and purpose		
In a day how long you work and how long the leisure period		
If you earn then can you expend independently if yes at what percent		
What are the agricultural activities the woman participate (name)		
Compare to 2009 do you think that your role in decision making has been changed (Greater involvement-1, same-2, less involvement-3)		
When do you take dinner and lunch (before my husband-1, same time with my husband, after my husband-3)		
Please comment about your daily food intake compare to adult male HH? (Higher-1, same-2, Less-3)		
Decision code	Main male or husband ……………1 Main female or wife ………………..2 Husband and wife jointly ………..3 Someone else in the household ….4 Jointly with someone else inside the household…..5 Jointly with someone else outside the household ….6 Someone outside the household/others7 Decision not made ………………..99	
Extent of participation in decision making	Not at all ……………………1 Small extent………………..2 Medium extent……………3 To a high extent…………4	

Place of visit (frequency)	Week	Month	Year
Market			
Relative house (outside village)			
Own upazila sadar			
Other upazila			
Own district sadar			
Other district			
Fisheries office			
Capital city			

29. Climate change perception
30.1 Have you heard or read about climate change? 1=Yes 2=No
If yes, which source?
1=TV 2=Radio 3=Newspaper 4=Bulletin 5=Extension agent 6=Other (specify)
What was said?

30.2 Have you being getting information on weather or climate? 1=Yes 2=No
If yes, from which source?
1=Radio 2=TV 3=Extension officers 4=Newspaper 5=Other (specify)
30.3 In your opinion, what do you think is the cause of the changes in rainfall, temperature etc
30.4 What are the perceived problems occurred due to change in temperature and precipitation (code)
1=Problems with drinking water, 2= Cannot cultivate the crops in due time, 3=Growth of crops has decreased, 4=Yield of crops has decreased, 5=Cannot go outside of house due to extreme heat, 6= Have to work hard for irrigation, 7= Diseases/health problems/sickness has increased, 8= Boro (summer paddy) cannot be cultivated timely, 9= Boro (summer paddy) seedbed cannot be sown, 10=Potato cultivation is hampered, 11= Betel leaf field is hampered, 12= Robi (winter crops) crops cannot be cultivated, 13= Potato cultivation is hampered because of dense fog, 14= Flowering/blooming is delayed, 15= Color of the crops has faded, 16= Mango inflorescence is hampered due to heavy fog, 17=others (specify)

END. Please thank your respondents for their time, and say that this will be used for PHD research, not for any other purpose.

Abstract

Integrated aquaculture-agriculture (IAA) is not a new or distinctive feature of farm households in Asia, however, innovations in IAA may make it a potentially sustainable intensification option for poor smallholder farmers in Asia, including Bangladesh. Taking this into consideration, this study hypothesised that relative to alternative options (e.g. rice monoculture) IAA has a potential role in improving livelihoods and poverty reduction among marginalized poor indigenous peoples of Bangladesh. IAA is a 'system technology' that potentially could be adopted by conventional rice, fish, poultry and livestock producers. The study examined the dynamics of IAA value chain participation by using a combination of three-year (2007, 2009 and 2012) panel data and one-year cross-sectional survey (2012) data from indigenous households in the northern and north-western regions of Bangladesh.

Based on the cross-sectional survey data from primary actors along rice–fish based IAA value chains, value chain mapping showed that there is little processing (mainly icing, grading and transportation) by value chain actors and that harvested fish are typically delivered to market or end consumers within a very short period (typically the same day). The quantitative results of gross margin, partial budgeting, and gender-based employment analyses indicate that replacing rice monoculture systems with rice–fish based IAA systems would likely result in positive socio-economic impacts. A strengths, weaknesses, opportunities, and threats (SWOT) analysis combined with value chain and partial budgeting analyses were applied to identify the technology, policy, and institutional level barriers that should be taken into account for realizing the full potential of rice–fish based IAA value chain development.

Technology adoption is a dynamic process, but the underlying dynamics have seldomly been studied using empirical approaches. In contrast, we examined IAA value chain participation dynamics and the factors that distinguish among non-participators, continuous participators, and those who begin but later discontinue participation (dis-participators) using different panel estimation methods (e.g. multinomial logit, random effects, and correlated random effects logit regression analyses) to control for omitted variables and endogenous regressors that very often are not possible to identify or may be ambiguous from cross-sectional studies. The results are consistent with determinants cited in the literature; more educated and larger households with better access to extension services and market information that participate in community-based organisations (CBO) are more likely to participate and continue participating in IAA value chain activities. Importantly, farm size and farm income did not appear to be positive significant determinants of IAA value chain participation, suggesting that IAA value chain activities are appropriate for resource-poor households. The results also indicate that distinct factors were associated with continuous participation and dis-participation in IAA value chain activities, especially for actors involved in upstream and downstream value chain activities. Important determinants of dis-participation were the age of household heads, the number of assets, farm size, fisheries income, and non-farm income.

After controlling for endogeneity of IAA value chain participation and unobserved heterogeneity using standard fixed effects and Heckit panel models, as well as control function approaches with three-wave panel data, the results indicated that IAA value chain participation was positively correlated to household income and consumption frequency of some goods, particularly fish, and that the benefits of participation did not continue to accrue after discontinuing participation in IAA value chains. We found evidence that IAA value chain participation had greater welfare impacts on relatively wealthier households that participated in production related activities than on landless, extremely poor households that participated in upstream and downstream value chain activities.

Assessment of the comparative socio-environmental impacts of rice monoculture and rice–fish based IAA practices based on cross-sectional survey data on plot level inputs and outputs, and farmer perceptions suggests that rice–fish based IAA is a sustainable alternative to rice monoculture. Although farmers were well aware of the more negative socio-environmental impacts of rice monoculture compared to rice–fish based IAA systems, their perceptions were limited to visible impacts such as the incidence of disease and crop pests, changes in soil structure, the relative quantities of beneficial organisms, soil fertility, fish harvests, etc. The results of the Tobit regression and propensity score matching (PSM) methods showed that the adoption of rice–fish based IAA systems was positively associated with farmer awareness of the negative impacts of rice monoculture. Promotion of the 'farmer field school' type institutional training approaches and infrastructure development may also play positive roles in raising awareness about the adverse socio-environmental impacts of rice monoculture. Overall the study results demand broader attention. A key general message is that, although the results reveal that the gains from IAA value chain participation remain substantial, many smallholders were not able to participate in IAA value chains or could not sustain their participation due to various factors.

Zusammenfassung

Integrierte Produktionssysteme aus Aquakultur und Landwirtschaft (IAA) sind keine neuen oder besonderen Merkmale von bäuerlichen Haushalten in Asien. Innovationen der IAA bieten jedoch eine nachhaltige Option zur Intensivierung der Landwirtschaft für arme Kleinbauern in Asien, darunter auch Bangladesch. Unter Berücksichtigung dieser Tatsachen stellt diese Studie die These auf, dass IAAs verglichen mit anderen Alternativen (z.b. Reismonokultur) das Potential besitzen, die Lebensumstände der marginalisierten, indigenen und armen Bevölkerung in Bangladesch zu verbessern und Armut zu bekämpfen. IAA als ‚Systemtechnologie' kann von konventionellen Reis-, Fisch-, Geflügel- und Nutzviehproduzenten angewandt werden. Diese Studie untersucht die Dynamiken der Partizipation in IAA-Wertschöpfungsketten mit einer Kombination aus Paneldaten, die in drei Zeitabschnitten erhoben wurden (2007, 2008 und 2012), und Querschnittsdaten für das Jahr 2012 von indigenen Haushalten im Norden und Nordwesten von Bangladesch.

Auf Grundlage detaillierter Querschnittsdaten der Hauptakteure entlang der auf Reis-Fisch basierten IAA Wertschöpfungskette, wurde durch die Nachzeichnung der Wertschöpfungskette aufgezeigt, dass nur eine geringe Verarbeitung durch die Akteure in dieser stattfindet. Hierzu zählen wozu hauptsächlich Vereisung, Qualitätseinstufung und Transport. Darüber hinaus konnte festgestellt werden, dass gefangener Fisch innerhalb eines sehr kurzen Zeitraums (üblicherweise am gleichen Tag) zum Markt oder dem Endverbraucher geliefert wird. Insgesamt weisen die quantitativen Ergebnisse bezüglich Bruttogewinn, Teilbudgetierung und der geschlechtsspezifischen Beschäftigungsanalyse darauf hin, dass das Ersetzen von Reismonokultursystemen durch Reis-Fisch basierte IAAs positive Effekte mit sich bringt. Eine SWOT (Stärke, Schwächen, Möglichkeiten und Gefahren)-Analyse in Kombination mit der Untersuchung von Wertschöpfungsprozessen und Teilbudgtierungen wurde angewandt, um die Technologie sowie politische und institutionelle Hürden zu identifizieren, die bei der bei der Entwicklung von Reis-Fisch basierten IAA Wertschöpfungsketten zur Ausnutzung des vollen Potenzials einbezogen werden sollten.

Dynamiken der technologischen Anpassung wurden bislang selten empirisch untersucht. Diese Studie analysierte die Teilnahmeentwicklungen an IAA Wertschöpfungsketten einschließlich der Unterschiede, die sich zwischen Nicht-Teilnehmern, beständigen Teilnehmern und denen, die mit der Teilnahme beginnen aber diese später unterbrechen, anhand unterschiedlicher Panelschätzmethoden (z.B. Multinomiale logistische Regression, Random-Effects-Modell und korrelierte logistische Random-Effects-Regressionsanalyse) erkennen lassen. Somit konnten fehlende und endogene Variablen kontrolliert werden, deren Identifizierung häufig nicht möglich oder deren Effekt in Querschnittsanalysen mehrdeutig ist. Die Ergebnisse stimmen mit den in der einschlägigen Literatur genannten Determinanten überein. Hierzu zählt, dass für besser ausgebildete, größere Haushalte, mit einem besseren Zugang zu Beratungsdiensten und Marktinformationen, in welchen die Mitglieder in gemeinschaftsbasierten Organisationen sind, die Wahrscheinlichkeit der Teilnahme und Fortführung an den Aktivitäten der IAA Wertschöpfungsketten höher ist. Wichtig ist, dass sich Betriebsgröße und landwirtschaftliches Einkommen nicht als positive, signifikante Determinanten der IAA Wertschöpfungsteilnahme erwiesen haben. Diese Erkenntnis lässt annehmen, dass die Aktivitäten im Rahmen des IAA Wertschöpfungsprozesses für ressourcenarme Haushalte angemessen sind. Die Ergebnisse weisen zudem darauf hin, dass unterschiedliche Faktoren mit der beständigen Teilnahme und dem Abbruch der Teilnahme an IAA-Wertschöpfungsprozessen zusammenhängen. Dies gilt insbesondere für Akteure, die in vor- und nachgeschaltete Wertschöpfungsaktivitäten einbezogen sind. Wichtige Determinanten der Nicht-Teilnahme sind das Alter des Haushaltsvortandes, die Anzahl an Besitztümern, die Größe des Betriebes sowie die Einnahmen durch Fischerei und nicht-landwirtschaftliche Tätigkeiten.

Die Endogenität der IAA-Teilnahme und die unbeobachteten Heterogenität wurden durch die Anwendung des Standard Fixed Effect-Modells, des Heckit Panel Modells und Kontrollfunktionsmethoden mit drei-periodigen Paneldaten kontrolliert. Die Ergebnisse weisen auf eine positive Korrelation der Beteiligung an IAA-Wertschöpfungsprozessen mit Haushaltseinkommen und der Konsumhäufigkeit verschiedener Güter (insbesondere Fisch) hin. Die Vorteile der Teilnahme wachsen jedoch nicht an nach dem Abbruch der Partizipation an IAA-Wertschöpfungsprozessen.

Die Bewertung der komparativen sozio-ökonomischen Auswirkungen von Reismonokultur und Reis-Fisch basierten IAA Praktiken durch Nutzung von Querschnittsdaten auf flächenbezogene Inputs und Erträge sowie Einblicke in die Sichtweise der Bauern legen nahe, dass Reis-Fisch basierte IAAs eine nachhaltige Alternative zur Reismonokultur darstellen. Auch wenn Bauern über die negativen sozio-ökonomischen Effekte der Reismonokultur im Vergleich zu Reis-Fisch basierten IAAs gut informiert sind, sind deren Einschätzungen auf sichtbare Auswirkungen wie etwa das Auftreten von Krankheiten und Schädlingen, Veränderungen im Grad der Bodenverdichtung, der relativen Menge an nützlichen Organismen, der Bodenfruchtbarkeit und der Menge des Fischfangs usw. begrenzt. Die Ergebnisse heben zusätzlich die Erkenntnisse der Tobit Regression und der

Propensity Score Matching (PSM) Methode hervor. Diesen zufolge hilft die Anwendung von Reis-Fisch basierten IAAs dabei das Bewusstsein der Bauern für negative Auswirkungen der Reismonokultur zu verbessern. Die Förderung von institutionellen Trainingsmethoden in Form von Schulungen, sowiedie Weiterentwicklung der Infrastruktur werden ebenfalls eine positive Rolle bei der Wahrnehmung negativer sozio-ökonomischer Auswirkungen der Reismonokultur haben. Insgesamt erfordern die Untersuchungsergebnisse weitere Aufmerksamkeit in der Zukunft. Eine zentrale Aussage ist, dass es vielen Kleinbauern nicht möglich ist, an der IAA Wertschöpfungskette beteiligt zu sein oder die Anwendung fortzuführen, obwohl die Ergebnisse wesentliche Gewinne durch die Anwendung von IAA-Wertschöpfungskettenaktivitäten offenbaren.

Development Economics and Policy

Series edited by Franz Heidhues†, Joachim von Braun, Ulrike Grote and Manfred Zeller

Vol. 1 Andrea Fadani: Agricultural Price Policy and Export and Food Production in Cameroon. A Farming Systems Analysis of Pricing Policies. The Case of Coffee-Based Farming Systems. 1999.

Vol. 2 Heike Michelsen: Auswirkungen der Währungsunion auf den Strukturanpassungsprozeß der Länder der afrikanischen Franc-Zone. 1995.

Vol. 3 Stephan Bea: Direktinvestitionen in Entwicklungsländern. Auswirkungen von Stabilisierungsmaßnahmen und Strukturreformen in Mexiko. 1995.

Vol. 4 Franz Heidhues / François Kamajou: Agricultural Policy Analysis – Proceedings of an International Seminar, held at the University of Dschang, Cameroon on May 26 and 27 1994, funded by the European Union under the Science and Technology Program (STD). 1996.

Vol. 5 Elke M. Förster: Protection or Liberalization? A Policy Analysis of the Korean Beef Sector. 1996.

Vol. 6 Gertrud Schrieder: The Role of Rural Finance for Food Security of the Poor in Cameroon. 1996.

Vol. 7 Nestor R. Ahoyo Adjovi: Economie des Systèmes de Production intégrant la Culture de Riz au Sud du Bénin: Potentialités, Contraintes et Perspectives. 1996.

Vol. 8 Jenny Müller: Income Distribution in the Agricultural Sector of Thailand. Empirical Analysis and Policy Options. 1996.

Vol. 9 Michael Brüntrup: Agricultural Price Policy and its Impact on Production, Income, Employment and the Adoption of Innovations. A Farming Systems Based Analysis of Cotton Policy in Northern Benin. 1997.

Vol. 10 Justin Bomda: Déterminants de l'Epargne et du Crédit, et leurs Implications pour le Développement du Système Financier Rural au Cameroun. 1998.

Vol. 11 John M. Msuya: Nutrition Improvement Projects in Tanzania: Implementation, Determinants of Performance, and Policy Implications. 1998.

Vol. 12 Andreas Neef: Auswirkungen von Bodenrechtswandel auf Ressourcennutzung und wirtschaftliches Verhalten von Kleinbauern in Niger und Benin. 1999.

Vol. 13 Susanna Wolf (ed.): The Future of EU-ACP Relations. 1999.

Vol. 14 Franz Heidhues / Gertrud Schrieder (eds.): Romania – Rural Finance in Transition Economies. 2000.

Vol. 15 Katinka Weinberger: Women's Participation. An Economic Analysis in Rural Chad and Pakistan. 2000.

Vol. 16 Christof Batzlen: Migration and Economic Development. Remittances and Investments in South Asia: A Case Study of Pakistan. 2000.

Vol. 17 Matin Qaim: Potential Impacts of Crop Biotechnology in Developing Countries. 2000.

Vol. 18 Jean Senahoun: Programmes d'ajustement structurel, sécurité alimentaire et durabilité agricole. Une approche d'analyse intégrée, appliquée au Bénin. 2001.

Vol. 19 Torsten Feldbrügge: Economics of Emergency Relief Management in Developing Countries. With Case Studies on Food Relief in Angola and Mozambique. 2001.

Vol. 20 Claudia Ringler: Optimal Allocation and Use of Water Resources in the Mekong River Basin: Multi-Country and Intersectoral Analyses. 2001.

Vol.	21	Arnim Kuhn: Handelskosten und regionale (Des-)Integration. Russlands Agrarmärkte in der Transformation. 2001.
Vol.	22	Ortrun Anne Gronski: Stock Markets and Economic Growth. Evidence from South Africa. 2001.
Vol.	23	Patrick Webb / Katinka Weinberger (eds.): Women Farmers. Enhancing Rights, Recognition and Productivity. 2001.
Vol.	24	Mingzhi Sheng: Lebensmittelkonsum und -konsumtrends in China. Eine empirische Analyse auf der Basis ökonometrischer Nachfragemodelle. 2002.
Vol.	25	Maria Iskandarani: Economics of Household Water Security in Jordan. 2002.
Vol.	26	Romeo Bertolini: Telecommunication Services in Sub-Saharan Africa. An Analysis of Access and Use in the Southern Volta Region in Ghana. 2002.
Vol.	27	Dietrich Müller-Falcke: Use and Impact of Information and Communication Technologies in Developing Countries' Small Businesses. Evidence from Indian Small Scale Industry. 2002.
Vol.	28	Wolfram Erhardt: Financial Markets for Small Enterprises in Urban and Rural Northern Thailand. Empirical Analysis on the Demand for and Supply of Financial Services, with Particular Emphasis on the Determinants of Credit Access and Borrower Transaction Costs. 2002.
Vol.	29	Wensheng Wang: The Impact of Information and Communication Technologies on Farm Households in China. 2002.
Vol.	30	Shyamal K. Chowdhury: Institutional and Welfare Aspects of the Provision and Use of Information and Communication Technologies in the Rural Areas of Bangladesh and Peru. 2002.
Vol.	31	Annette Luibrand: Transition in Vietnam. Impact of the Rural Reform Process on an Ethnic Minority. 2002.
Vol.	32	Felix Ankomah Asante: Economic Analysis of Decentralisation in Rural Ghana. 2003.
Vol.	33	Chodechai Suwanaporn: Determinants of Bank Lending in Thailand: An Empirical Examination for the Years 1992 to 1996. 2003.
Vol.	34	Abay Asfaw: Costs of Illness, Demand for Medical Care, and the Prospect of Community Health Insurance Schemes in the Rural Areas of Ethiopia. 2003.
Vol.	35	Gi-Soon Song: The Impact of Information and Communication Technologies (ICTs) on Rural Households. A Holistic Approach Applied to the Case of Lao People's Democratic Re- public. 2003.
Vol.	36	Daniela Lohlein: An Economic Analysis of Public Good Provision in Rural Russia. The Case of Education and Health Care. 2003.
Vol.	37	Johannes Woelcke. Bio-Economics of Sustainable Land Management in Uganda. 2003.
Vol.	38	Susanne M. Ziemek: The Economics of Volunteer Labor Supply. An Application to Countries of a Different Development Level. 2003.
Vol.	39	Doris Wiesmann: An International Nutrition Index. Concept and Analyses of Food Insecurity and Undernutrition at Country Levels. 2004.
Vol.	40	Isaac Osei-Akoto: The Economics of Rural Health Insurance. The Effects of Formal and Informal Risk-Sharing Schemes in Ghana. 2004.
Vol.	41	Yuansheng Jiang: Health Insurance Demand and Health Risk Management in Rural China. 2004.

Vol. 42 Roukayatou Zimmermann: Biotechnology and Value-added Traits in Food Crops: Relevance for Developing Countries and Economic Analyses. 2004.

Vol. 43 F. Markus Kaiser: Incentives in Community-based Health Insurance Schemes. 2004.

Vol. 44 Thomas Herzfeld: *Corruption begets Corruption*. Zur Dynamik und Persistenz der Korruption. 2004.

Vol. 45 Edilegnaw Wale Zegeye: The Economics of On-Farm Conservation of Crop Diversity in Ethiopia: Incentives, Attribute Preferences and Opportunity Costs of Maintaining Local Varieties of Crops. 2004.

Vol. 46 Adama Konseiga: Regional Integration Beyond the Traditional Trade Benefits: Labor Mobility contribution. The Case of Burkina Faso and Côte d'Ivoire. 2005.

Vol. 47 Beyene Tadesse Ferenji: The Impact of Policy Reform and Institutional Transformation on Agricultural Performance. An Economic Study of Ethiopian Agriculture. 2005.

Vol. 48 Sabine Daude: Agricultural Trade Liberalization in the WTO and Its Poverty Implications. A Study of Rural Households in Northern Vietnam. 2005.

Vol. 49 Kadir Osman Gyasi: Determinants of Success of Collective Action on Local Commons. An Empirical Analysis of Community-Based Irrigation Management in Northern Ghana. 2005.

Vol. 50 Borbala E. Balint: Determinants of Commercial Orientation and Sustainability of Agricultural Production of the Individual Farms in Romania. 2006.

Vol. 51 Pamela Marinda: Effects of Gender Inequality in Resource Ownership and Access on Household Welfare and Food Security in Kenya. A Case Study of West Pokot District. 2006.

Vol. 52 Charles Palmer: The Outcomes and their Determinants from Community-Company Contracting over Forest Use in Post-Decentralization Indonesia. 2006.

Vol. 53 Hardwick Tchale: Agricultural Policy and Soil Fertility Management in the Maize-based Smallholder Farming System in Malawi. 2006.

Vol. 54 John Kedi Mduma: Rural Off-Farm Employment and its Effects on Adoption of Labor Intensive Soil Conserving Measures in Tanzania. 2006.

Vol. 55 Mareike Meyn: The Impact of EU Free Trade Agreements on Economic Development and Regional Integration in Southern Africa. The Example of EU-SACU Trade Relations. 2006.

Vol. 56 Clemens Breisinger: Modelling Infrastructure Investments, Growth and Poverty Impact. A Two-Region Computable General Equilibrium Perspective on Vietnam. 2006.

Vol. 57 Meike Wollni: Coping with the Coffee Crisis. An Analysis of the Production and Marketing Performance of Coffee Farmers in Costa Rica. 2007.

Vol. 58 Franklin Simtowe: Performance and Impact of Microfinance. Evidence from Joint Liability Lending Programs in Malawi. 2008.

Vol. 59 Xiangping Jia: Credit Rationing and Institutional Constraint. Evidence from Rural China. 2008.

Vol. 60 Holger Seebens: The Economics of Gender and the Household in Developing Countries. 2009.

Vol. 61 Stephan Piotrowski: Land Property Rights and Natural Resource Use. An Analysis of Household Behavior in Rural China. 2009.

Vol. 62 Sebastian M. Scholz: Rural Development through Carbon Finance. Forestry Projects under the Clean Development Mechanism of the Kyoto Protocol. Assessing Smallholder Participation by Structural Equation Modeling. 2009.

Vol.	63	Jakob Rupert Friederichsen: Opening Up Knowledge Production through Participatory Research? Agricultural Research for Vietnam's Northern Uplands. 2009.
Vol.	64	Olivier Ecker: Economics of Micronutrient Malnutrition. The Demand for Nutrients in Sub-Saharan Africa. 2009.
Vol.	65	Julia Johannsen: Operational Assessment of Monetary Poverty by Proxy Means Tests. 2009
Vol.	66	Ephraim Nkonya / Nicolas Gerber / Philipp Baumgartner / Joachim von Braun / Alessandro De Pinto / Valerie Graw / Edward Kato / Julia Kloos / Teresa Walter: The Economics of Land Degradation. Toward an Integrated Global Assessment. 2011.
Vol.	67	S. Idriss Nazaire Houssou: Operational Poverty Targeting by Proxy Means Tests. Models and Policy Simulations for Malawi. 2013.
Vol.	68	Abdul Salam Lodhi: Education, Child Labor and Human Capital Formation in Selected Urban and Rural Settings of Pakistan. 2013.
Vol.	69	Evita Hanie Pangaribowo: Household Food Consumption, Women´s Asset and Food Policy in Indonesia. 2013.
Vol.	70	Dan Liu: China's New Rural Cooperative Medical Scheme. Evolution, Design and Impacts. 2013.
Vol.	71	Camille Saint-Macary: Microeconomic Impacts of Institutional Change in Vietnam's Northern Uplands. Empirical Studies on Social Capital, Land and Credit Institutions. 2014.
Vol.	72	Beatrice Wambui Muriithi: Commercialization of Smallholder Horticultural Farming in Kenya. Poverty, Gender, and Institutional Arrangements. 2014.
Vol.	73	Christian C. W. Grovermann: Assessment of Pesticide Use Reduction Strategies for Thai Highland Agriculture. Combining Econometrics and Agent-based Modelling. 2015.
Vol.	74	Dawit Diriba Guta: Bio-Based Energy, Rural Livelihoods and Energy Security in Ethiopia. 2015.
Vol.	75	Tigabu Degu Getahun: Industrial Clustering, Firm Performance and Employee Welfare. Evidence from the Shoe and Flower Cluster in Ethiopia. 2016.
Vol.	76	Abu Hayat Md. Saiful Islam: Impact of Technological Innovation on the Poor. Integrated Aquaculture-Agriculture in Bangladesh. 2016.

www.peterlang.com